Treatment and Disposal
of Pesticide Wastes

A C S   S Y M P O S I U M   S E R I E S **259**

Treatment and Disposal of Pesticide Wastes

Raymond F. Krueger, EDITOR
Environmental Protection Agency

James N. Seiber, EDITOR
University of California, Davis

Based on a symposium sponsored by
the Division of Pesticide Chemistry
at the 186th Meeting
of the American Chemical Society,
Washington, D.C.,
August 28–September 2, 1983

American Chemical Society, Washington, D.C. 1984

Library of Congress Cataloging in Publication Data

Treatment and disposal of pesticide wastes.
  (ACS symposium series, ISSN 0097–6156; 259)

  Includes bibliographies and indexes.

  1. Pesticides industry—Waste disposal—Congresses.

  I. Krueger, Raymond F. II. Seiber, James N.,
1940–   . III. American Chemical Society. Division of
Pesticide Chemistry.

TD899.P37T74   1984   668′.65   84–12327
ISBN 0–8412–0858–1

ACS Symposium Series

M. Joan Comstock, *Series Editor*

Advisory Board

FOREWORD

The ACS SYMPOSIUM SERIES was founded in 1974 to provide a medium for publishing symposia quickly in book form. The format of the Series parallels that of the continuing ADVANCES IN CHEMISTRY SERIES except that in order to save time the papers are not typeset but are reproduced as they are submitted by the authors in camera-ready form. Papers are reviewed under the supervision of the Editors with the assistance of the Series Advisory Board and are selected to maintain the integrity of the symposia; however, verbatim reproductions of previously published papers are not accepted. Both reviews and reports of research are acceptable since symposia may embrace both types of presentation.

CONTENTS

PREFACE

IMPROPER HANDLING AND DISPOSAL of chemical wastes represents a serious problem in the United States and in other parts of the world. Although the recent spotlight has been on the management of hazardous industrial wastes and the paucity of properly designed facilities for their treatment, storage, or disposal, a keen interest also exists in developing better disposal strategies for pesticide containing wastes. Continuing emphasis on greater crop production and the resulting need to control pests, often generally accomplished by the use of pesticide chemicals in large quantities, has created a problem of waste disposal. Because of the quantities used, areas affected, and potential hazards, pesticides must rank high in priority for the development of improved waste disposal practices.

Several studies have been conducted during the last five or six years that bear directly on pesticide waste disposal, but many have yet to be published in the open literature. Some of these studies were just being initiated when ACS SYMPOSIUM SERIES NO. 73 was published. In addition, federal requirements for pesticide disposal under the Resource Conservation and Recovery Act (RCRA) and the Federal Insecticide, Fungicide, and Rodenticide Act (FIFRA) as amended have recently been modified and/or clarified to provide more explicit rules and guidance to those with waste disposal needs. For these reasons, a symposium was held by the American Chemical Society that brought together papers on the chemical, biological, and physical methods of disposal; the basic principles underlying each of these methods; the analytical and modeling approaches applicable to waste disposal conditions; and the EPA regulatory viewpoint on safe disposal. This symposium was the basis for this volume.

Special thanks go to the co-organizers of the symposium, D. Kaufmann, J. Plimmer, and D. Severn; to P. Roberts, who, through the ACS Division of Environmental Chemistry, cosponsored the symposium; and to the ACS Division of Pesticide Chemistry, which sponsored the symposium, arranged the scheduling, and provided some funding.

RAYMOND F. KRUEGER
Environmental Protection Agency—Washington

JAMES N. SEIBER
University of California, Davis

May 3, 1984

INTRODUCTION

At several stages in the lifetime of a given pesticide, wastes are generated that need attention. These stages include manufacturing; testing; formulation; transport; field mixing and application; the recall of cancelled, suspended, or outdated products; and the enormous numbers of empty containers left after use of the chemicals. Although none of these stages commonly generate large quantities of waste (except for the empty containers) the total amount of waste generated over the pesticide's useful lifetime can be substantial.

The pesticide wastewaters generated by commercial aerial applicators is just one example of the magnitude of the disposal problem. Commercial aerial applicators account for roughly half of all pesticide applications in the United States. Every day, each sprayplane can generate 30 gallons of aqueous rinsate, principally from washing residual material from the hoppers, containing approximately 500 ppm of residue (1,2)—the simple act of washing the exterior of the aircraft makes a contribution. By combining these figures with the number of aircraft involved in commercial spraying (6,000) one may calculate a potential volume of 60 million gallons containing 120,000 kg of chemical residue each year from this single source. Although often these solutions are legally used as spray diluent, a substantial proportion may be considered as waste in need of disposal.

The disposal of these pesticide-containing wastewaters varies considerably in terms of legal compliance and safety. In California, for example, more than 100 haphazard wastewater dumpsites are believed to exist in the Central Valley; a few of these sites have been highlighted by recent adverse publicity. Wells at applicators' facilities have been found to be contaminated as a result of the mismanagement of these wastes. On the other hand, the University of California has maintained soil pit disposal beds at its field stations just for handling pesticide wastewater, with generally favorable results. Comparable results have been experienced by others using similar facilities. According to a recent survey of commercial aerial applicators across the U.S., a large number wanted to improve their wastewater systems, if acceptable methods were made known to them (3).

This one area—disposal of wastewater by commercial applicators—demonstrates the need that exists not only for better disposal strategies, but also for the more effective dissemination of information relating to them. The chapters in this book provide a source of information to attain these goals.

Literature Cited

1. "Disposal of Dilute Pesticide Solutions," SCS Engineers, 1979, NTIS Report PB-297 985.
2. Whittaker, K. F.; Nye, J. C.; Wukash, R. F.; Squires, R. G.; York, N. C.; and Kazimier, H. A. "Collection and Treatment of Wastewater Generated by Pesticide Applicators." Unpublished report, Purdue University, West Lafayette, Indiana, 1979.
3. Craigmill, A. C.; and Seiber, J. N. Unpublished data.

FEDERAL REGULATIONS

Regulation of Pesticide Disposal

RAYMOND F. KRUEGER and DAVID J. SEVERN

Office of Pesticide Programs, Environmental Protection Agency, Washington, DC 20460

One result of the annual use of millions of pounds of pesticide chemicals each year is the production of numerous empty containers and other pesticide wastes. When the potentially toxic characteristics of the many pesticide formulations that are used are taken into consideration, the possibility of injury to man and the environment due to improper disposal can be considerable. Information on improved disposal technologies has been published to provide better management of pesticide wastes but fall short in that they do not cover all wastes or disposal methods. Regulations that are intended to control improper disposal have been promulgated under FIFRA and RCRA which, in general, are intended to control disposal methods in common use today such as land disposal, incineration, open burning, certain physical/chemical methods, and some systems that utilize biological degradation. The RCRA regulations also provide standards for construction and operation of certain disposal facilities. The regulations do not provide specific information to the pesticide user as to how to dispose of his wastes. One way to make such information readily available is to put it on the label of each pesticide product. Guidelines that establish data requirements to register certain pesticides have been published. Similar guidelines for disposal statements are being prepared the Environmental Protection Agency.

History of Disposal Regulation

In the late 1960's, the seriousness of the hazard to human health and the environment resulting from mismanagement of pesticide wastes in particular and hazardous wastes in general became in-

creasingly clear. As a part of this general awareness, and as
the use of pesticides continued to grow, resulting in more and
more empty containers and other wastes to be disposed of, the
problem of pesticide waste disposal became a point of major
concern. This focus of interest was heightened by an increase in
documented cases of pesticide disposal mismanagement such as
injuries to children playing with empty pesticide containers or
fish kills resulting from so-called empty containers being care-
lessly dumped into streams or ponds (1).

In 1972 Congress enacted the Federal Environmental Pesticide Control Act (FEPCA) which amended the Federal Insecticide,
Fungicide and Rodenticide Act (FIFRA) to include, among other
additions, attention to the pesticide disposal problem. Specifically, Section 19 was added which says, "The Administrator
shall...establish procedures and regulations for the disposal or
storage of packages and containers of pesticides, and for disposal
or storage of excess amounts of such pesticides, and accept at
convenient locations for safe disposal a pesticide the registration of which is cancelled under section 6(c) if requested by the
owner of the pesticide." Section 19 was further modified in 1978
to require information on disposal to accompany all cancellation
orders.

On May 1, 1974, in response to the mandate for procedures
and regulations, the Agency published in the Federal Register,
"Pesticides and Pesticide Containers, Regulations for the Acceptance and Recommended Procedures for Disposal and Storage of
Pesticides and Pesticide Containers" (40 CFR 165) (2). The
Regulatory portion makes up a relatively small part of the total
package, prescribing the process owners of pesticides the regis-
trations of which have been suspended and cancelled, must follow
in order for the EPA to accept the products that qualify for
disposal. The remaining parts of the May 1, 1974, publication
provide general guidance on disposal of pesticides and of empty
containers. These are recommendations only and have no force or
effect of law. The basic objectives of the publication were to
meet the requirements of the law, to provide agency policy on
pesticide disposal and to give guidance to the states where none
had existed before.
Another piece of legislation having impact on the hazardous waste
problem is the Resource Conservation and Recovery Act (RCRA).
Passage of this act in 1976 was stimulated by several episodes of
severe mismanagement of hazardous wastes that received considerable attention in the news media. RCRA is intended to provide
control over management of hazardous wastes from point of generation, through transport and to final treatment, storage or disposal. This "cradle-to-grave" coverage has had an impact on
pesticide disposal activities. However, as will be explained
later, the regulations promulgated under RCRA do not cover all
pesticides.

Current Disposal Practices

Pesticides have been used in large quantities for many years and
during that time the wastes that have been generated were disposed of in various ways. Past practices have not necessarily
been satisfactory, as previously noted; however, it is also true
that some of the disposal methods that were used fifty years ago
are still in use today.
        According to a USDA survey done in 1972 (3), initial dispo-
sition of empty pesticide containers was done as follows:

Returned to dealer	3.1%
Burned	49.2%
Buried	5.8%
Private Dump	18.9%
Commercial Dump	0.9%
Left in field	1.1%
Left where sprayer filled	8.4%
Retained	11.0%
Other	1.6%
	100.0%

The method of container disposal given does not mean final "rest-
ing place" in the environment of any of the residual material
still in the container. It only indicates what farmers did with
the container initially.
        Field surveys conducted as a part of an economic analysis
of pesticide disposal in California, Iowa, New York and Mississippi (4) showed a fairly consistent pattern. In California,
combustible containers, paper, plastic, etc., were generally
burned on-site. Metal and glass containers were disposed of in
sanitary landfills. Rinsing of empty containers, which is
required under some circumstances, was found to be a common
practice. A similar pattern was found in Iowa and New York.
Mississippi differed only in that the state operates a rural
collection system. Trash collection containers are positioned at
strategic locations to accept empty containers. The contents are
periodically collected and delivered to sanitary landfills that
have been specially designated by the State Health Department.
Farmers are urged to rinse all containers prior to putting them
into the collection system for disposal. The study also noted
that pesticides are handled in bulk or in 55 gallon drums more
frequently in the South than in other parts of the country. The
empty 55 gallon drums are generally made available to a drum
reconditioner who collects them at regular intervals.
        As a practical matter, disposal is tied closely with econo-
mics. For example, empty 55 gallon drums have value to drum
reconditioners who frequently seek them out to recondition for
reuse. Also, during the mid-1970's when the price of scrap steel

soared to unusual heights, empty five gallon cans were collected
and sold for scrap. When the price of scrap steel dropped, the
practice ended. In one case, about a quarter of a million cans
that had been collected for scrap were abandoned. The pile stood
for several years as a monument to the volume of pesticides used
in that area.

Although "high-tech" solutions to disposal are readily
available they are not put to use because of the cost factor.
The more common disposal methods are those that have been with us
for some time.

EPA Guidance on Disposal

EPA is looked to for guidance in the field of pesticide disposal;
however, in the past a low priority was generally assigned to
development of waste management facilities and systems. Minimal
resources were allocated to the development of new disposal
technologies. To be required to spend money to throw something
away is not an acceptable situation. Thus users of pesticides
are inclined to look for inexpensive ways of getting rid of
pesticide wastes. Strategies for waste management do exist, but
are often too expensive to be readily accepted. Simple, inexpensive disposal systems are not generally available, largely because there is no one method that can be safely employed in every
situation.

The complexity of choosing a disposal method or providing
safe guidance for the disposal of a particular product becomes
apparent when some of the factors that must be considered in
evaluating a disposal action are listed:

-Chemical, physical, biological and toxicological characteris-
 tics of the formulated products;
-Composition, concentration and quantity of the waste;
-Size, composition and numbers of waste containers;
-Geographic location of the wastes;
-Availability, utility and relative costs of disposal methods
 and facilities;
-Technical and economic feasibility of recycling the wastes;
-Attitude of the pesticide industry, user, and the general
 public concerning waste disposal;
-State regulatory requirements that affect the proposed dis-
 posal action, such as deposit/return laws.

Given this complex maze of information needs, the EPA is
expected to provide guidance on disposal of pesticide wastes in
all parts of the continental United States as well as in other
parts of the world. This has created a serious challenge.

Different options for getting disposal information to the pesticide user have been tried. For example, a number of manuals were published to provide advice on specific disposal problems. These were intended to provide "how-to" information to the people in the field who advise and regulate pesticide users. Each made a definite contribution, but information gaps remained.

"Guidelines for the Disposal of Small Quantities of Unused Pesticides" (5), published in 1975, was and EPA sponsored study by Midwest Research Institute. Most of the pesticides that were in common use at the time were reviewed and grouped by chemical class. The report gave disposal technologies that were described in the literature. Although the report brought together a wealth of information on chemical characteristics such as toxicity, solubility, and volatility, actual disposal advice was limited.

Another publication, "Handbook for Pesticide Disposal by Common Chemical Methods" (6), was also published in 1975 by EPA. This report evaluated 20 common pesticides and concluded that only 7 could be disposed of by alkali and/or acid hydrolysis. A detailed procedure for disposal by hydrolysis of the specific pesticides was also given. However, again, the information was of limited utility. A follow-up report covered another forty pesticides (7).

An EPA report "Disposal of Dilute Pesticide Solutions" (8) summarized technologies used in disposal of such wastes. Another study entitled "Economic Analysis of Pesticide Disposal Methods" (4) evaluated commonly used disposal systems in terms of cost. Local, regional and statewide collection and disposal strategies were also considered. Although each of these reports contain valuable information, they fall short of the target of providing adequate information to pesticide regulators and users to draw upon in day-today operations.

Another approach to developing information was to study specific technologies with an eye toward development of selected disposal methods. Some of the systems that were studied or are being studied include incineration, a catalytic dechlorination system utilizing nickel boride, ozone/UV radiation, and microwave plasma destruction. A summary of the research being done at the time was published in August 1978 under the title "State of the Art Report: Pesticide Disposal Research" (9). While much of this work has produced needed research information, the problem of making practical information available to the user remains. At the same time, an increasing public awareness of the problem was forcing a decision as to how the responsibility for providing disposal information would be handled.

RCRA/FIFRA

Although regulations covering broad areas of hazardous waste treatment, storage and disposal have been promulgated, the disposal of pesticide wastes is not totally under RCRA control. There are several reasons for this. Under RCRA, pesticide wastes are treated as any another waste and many fail to qualify as hazardous under the standards set forth in the regulations. This could be due to a combination of small amounts and low toxicity. For example, 2,4-D or DDT wastes in quantities of less that 1,000 kilograms per month would not be regulated under RCRA. Pesticide products registered for household use are simply not addressed by RCRA and are therefore not regulated. Empty containers are not regulated as "hazardous wastes" under RCRA; however, for certain highly toxic wastes, including some pesticides, the container must be triple rinsed (or given an equivalent treatment) before it is considered "empty" and thus not subject to control regulations (40 CFR 261.7). Also, in order to reduce the enormous regulatory workload promised by the burden of enforcing the regulations, a special exemption for farmers was written into the regulations (40 CFR 262.51). This exemption provides that a farmer who triple rinses his empty containers and disposes of them on his own property will be exempt from the requirements of RCRA. Commercial applicators do not enjoy such an exemption.

Given the complexity of the problem of selecting a safe, effective waste disposal strategy from the relatively sophisticated array of disposal systems that are available, and given the RCRA farmer's exemption, there needs to be a way of providing disposal information directly to the pesticide user. One such method would be to put it on the label.

Why put it on the label? One reason is that it would make compliance with the requirements of the disposal statement mandatory. Section 12 (a)(2) (G) of FIFRA says it is unlawful to use any pesticide in a manner inconsistent with its labeling, and disposal has been determined to be part of the use process. This means that the a label disposal statement would be, in effect, a regulation. Another reason is that the disposal directions can be tailored to fit the entire package, the container, the chemical it contains, and the site and mode of use. An appropriate disposal statement can be extremely important as one pesticide manufacturer (registrant) found out. There is a story circulated in EPA concerning the registrant who submitted a special label for approval. The comments contained in the response from EPA granting approval advised the registrant to include a statement on the label to provide guidance for disposal of the empty container which read: "Crush and bury, do not re-use". The registrant was heard to complain that he had intended to use the label on a railroad tank car.

The Label Statement and Labeling

Placing the disposal statement on the label means that development of the label statement becomes the responsibility of the
registrant who created the product in the first place. Given his
knowledge of the characteristics of the chemical formulation, he
is in the best position to provide an environmentally safe disposal method. Indeed, he is offered the opportunity to use a
broad range of sophisticated technologies rather than the famil-
iar "Crush and Bury, Do Not Reuse". This could assume considerable importance in the future given changing container economics
and disposal costs. Since the label accompanies the product,
necessary and specific information is placed directly in the
hands of the user. This is particularly important for farmers if
they wish to avail themselves of the exemption provided by RCRA
to avoid the extensive responsibilities attached to being a
"generator" of hazardous waste.
        In the foregoing discussion as well as what follows, the
term "label" is not limited to the printed matter attached to the
container. More correctly, the term "labeling" should be used
which refers to the printed information on the container as well
as any information that may accompany or refer to the product such
as pamphlets, books, or other printed material even though it may
not be physically attached to the container or in close proximity.
It is possible that relatively long, detailed disposal instructions could be developed and included in the labeling, but as a
practical matter, only guidance on disposal of the empty container
will appear on the package. Information on disposal of left-over
tank mixes, dilute solutions or unwanted product would be supplied
in the labeling or accompanying literature. In this way relatively
complex disposal procedures, complete with safety instructions,
can be provided to the user of the pesticide.

FIFRA Regulatory Requirements

The regulations for labeling pesticides are found in 40 CFR
162.10. These rules require that disposal information be a part
of any proposed statement. The current instructions to registrants on disposal statements are contained in PR Notice 83-3
(Pesticide Registration Notice 83-3). This notice advises registrants that the labels of all products must contain updated
disposal statements. The specific statements that are provided
cover most types of products such as home and garden or agricultural and most of the containers, metal, plastic or paper. A
typical recommended statement, such as one for metal containers,
reads as follows: "Triple rinse (or equivalent). Then offer for
recycling or reconditioning, or puncture and dispose of in a
sanitary landfill, or by other procedures approved by state and
local authorities".

The Office of Pesticide Programs recognizes that the disposal statements in this notice may not be appropriate for every
pesticide. Registrants have the option of proposing alternative
language for pesticide disposal statements. Registrants proposing alternate language must submit proposals to EPA and receive
approval before using any alternative language.

Under FIFRA regulations registered products must have a
disposal statement that is set apart and clearly distinguishable
from other directions for use. The regulations also require that
the disposal instructions be grouped together and printed in a
specified type size, depending on the size of the label front
panel. PR notice 83-3 is an attempt to improve and standardize
the label statements on disposal. Although the statements provided meet the regulatory need, they clearly fall short of providing the technical direction to the user that is required to
assure a high degree of environmental protection. For example,
minimal guidance is provided to the farmer that wants to take
advantage of the RCRA exemption and dispose on his own properly.
Although FIFRA regulations require data to support proposed label
statements on disposal, the statements are now provided by EPA
and such data is not required. Clearly, before a data requirement could be imposed, the registrant must be thoroughly advised
as to what data he must submit. This is the objective of the
Registration Guidelines.

Registration Guidelines

A considerable history is attached to regulation of the present
registration process as regards pesticide disposal statements.
On July 3, 1975, the Agency promulgated final regulations, 40 CFR
part 163, Subpart A (10). These regulations established the
basic requirements for registration of pesticide products.

During the period extending from 1975 to 1981, EPA issued
or made available several subparts of the guidelines for register-
ing pesticides in the United States which described, with more
specificity, the kinds of data that must be submitted to satisfy
the requirements of the registration regulations. These guidelines included sections detailing what data are required and
when, the standards for conducting acceptable tests, guidance on
the evaluation and reporting of data, and examples of acceptable
protocols.

In October of 1981, EPA decided to reorganize the guidelines and limit the regulation to a concise presentation of the
data requirements and when they are required. Therefore, data
requirements for pesticide registration pertaining to all former
subparts of the guidelines are now specified in part 158 (40 FR
53192 November 24, 1982) which specifies the kinds of data and

information that must be submitted to EPA to support the registration of each pesticide under the FIFRA. The standards for
conducting acceptable tests, guidance on evaluation and reporting
of data, further guidance on when data are required and examples
of protocols are not specified in part 158. This information
constitutes the guidelines and is available as advisory documentation through the National Technical Information Service (NTIS).
Guidelines were published in 1982 under such titles as: Subdivision D Chemistry Requirements: Product Chemistry; Subdivision
F Hazard Evaluation: Humans and Domestic Animals; Subpart G
Product Performance; Subdivision N Chemistry Requirements:
Environmental Fate, among others. Subdivision P: Disposal Data
Requirements was reserved for future publication.

The basic purpose of each of these guidelines is to provide
EPA with data to evaluate:
1. Direct hazard to humans and domestic animals;
2. Direct hazard to fish and wildlife;
3. Potential for contaminating ground water;
4. Potential for magnification in the food chain;
5. Potential uptake by rotational crops.

Specific data requirements are based on use patterns and
are listed in Section 158. For example, in Subdivision N: Environmental Fate, use patterns fall into the categories of terrestrial uses, aquatic and aquatic impact uses. Terrestrial uses
include domestic outdoor, green house, non-crop, orchard crop,
etc., and data required varies with the use site. Depending on
the site of use, studies on degradation, metabolism, mobility,
dissipation, and accumulation might be required. The general
guideline format is as follows:

a. Purpose
b. When required
c. Test standards
   (1.) test substances
   (2.) test procedures
d. Reporting and evaluation
e. References

The Subpart N, Environmental Fate Guidelines (11) will
probably provide much of what is needed to support many of the
disposal statements that would probably be proposed. However,
many of the studies would have to be conducted at "disposal
rates", that is hundreds of pounds or possibly tons per acre as
opposed to "use rates" of ounces or a few pounds per acre.

Disposal Technologies and Data Requirements

Based on use patterns and site-of-use, specific data are required
by the Section 158 regulations for each product registered.
Depending on the disposal technology selected, the information
needed to support the label disposal statement may well be additional to that required by Section 158. The following discussion
covers a few of the many disposal systems that might be candidates
for label statements and some of the kinds of data that might be
required.

Land disposal. Land disposal is the most widely used, least expensive, most often available disposal system at the present
time. The term land disposal includes sanitary landfills, surface impoundments, evaporation ponds and land farming. Land
disposal in a sanitary landfill, specially permitted to accept
such wastes, can be expected to be the method of choice for the
majority of the label statements proposed. Empty containers,
waste pesticides and other wastes are commonly disposed of in a
sanitary landfill or buried at the site of use.
      The soil is a complex and highly variable mixture of compo-
nents, containing many types of living organisms bacteria, fungi,
algae, invertebrate animals — and supporting the life of higher
plants, invertebrate and vertebrate animals. The addition of a
pesticide to a soil may therefore have effects on many living
organisms, and may in turn be affected by them. The pesticide
is also affected by the nature of the soil and by the climate
variables that affect the soil. Data on these factors and their
interrelationships must be developed before any land disposal
method that may impact those functions can be fully evaluated.
      An example of the kinds of data required for land disposal
options would be information on soil/pesticide interactions to
determine the effect of the pesticide on the soil and soil on the
pesticide. The physical composition of the soil and the physical
properties of the pesticide and its formulation will determine
the adsorption, leaching, water dispersal, and volatilization of
the pesticide which, in turn, determine the mobility of the
pesticide in soil. Even pesticides of closely related structures
may have very different soil retention properties. Much of this
data will be available from that developed to meet other registration data requirements with the exception that disposal rates are
often orders of magnitude higher than normal application rates
and the difference must be considered.
      Other considerations would include; data on adsorption and
leaching or other movement of the pesticide in the soil; the
effects of the pesticide on microorganisms in the soil under
aerobic and anaerobic conditions; effects of microorganisms on
the pesticide; effects of the pesticide on higher plants and

animals; effects on disposal facilities, such as liners or other structural material; and such other considerations that may be needed to determine if the proposed pesticide can be safely and effectively disposed of by land disposal means (12)(13). If the pesticide to be disposed of is specifically controlled under RCRA, then a very specific set of regulations is in force (40 CFR 261 to 270) that requires management in a permitted facility.

Incineration. A "pesticide incinerator" is defined as "any installation capable of the controlled combustion of pesticides, at a temperature of 1000` C (1832`F) for two seconds dwell time in the combustion zone, or lower temperatures and related dwell times that will assure complete conversion of the specific pesticide to inorganic gases and solid ash residues" (2). In addition, an incinerator must meet the performance standards promulgated under RCRA (40 CFR 264 Subpart O) if pesticides regulated under RCRA are to be burned. This means that an incinerator must be capable of destroying or removing 99.99% of the pesticide put into it. Test burns that are fully monitored are normally required to determine whether this performance standard is achieved. A registrant planning to suggest incineration on a disposal statement would need to observe these requirements.

Incineration of pesticides and/or containers requires special equipment that is not widely available. Due to the highly specialized nature of an incinerator that can meet the specifications necessary to destroy complex pesticide formulations, plus the energy requirements, the process can be very expensive and not generally the method of choice for small quantities that may be generated by a farmer, for example. On the other hand, it can be a highly effective means of disposing of unwanted material (14).

Open burning. "Open burning" is defined as combustion of a pesticide or pesticide container in any fashion other than incineration (2). Open burning is usually done by the simple act of piling up empty paper bags or plastic jugs and setting them on fire and is commonly used to dispose of combustible empty containers where local regulations permit the practice. It is sometimes prohibited by Regional Air Quality regulations. Where it is permitted open burning represents an inexpensive and convenient way of disposing of the combustible containers that are commonly used to package pesticides. The practice can, however, present hazards to worker health and to other persons, and to plants and animals that may be in the vicinity. The impact upon the environment is mainly through dispersal of combustion gases, smoke and fumes into the atmosphere and through contamination of soils and waters by ashes and partially burned containers holding toxic residues. Data would be required to address these issues.

Thermal degradation studies might be required to determine
the decomposition characteristics of of the subject pesticide
when heated alone or in the presence of oxidizers and/or binders
in both closed and open systems and at various temperatures.
Data required could cover the amounts and kinds of residuals that
might be found in the off-gases or remain in the ash. The igni-
tion characteristics of container materials and their maximum
burning temperatures must be tested under conditions simulating
the actual open burning conditions seen in the field. Data on
the results of burning small quantities of a subject pesticide in
a sample of the packaging material might also be required. Such
studies could be designed to show the composition of the end-
product gases produced by combustion at temperatures normally
achieved by burning wood, paper, cardboard, or plastics. Information on any special procedures that might be required or recommended would also be useful in determining if open burning of the
subject product would be safe.

Physical/chemical methods. Chemical deactivation/detoxification
provides the opportunity to reduce a toxic chemical to a non-toxic
state. It is a procedure that is not currently used to any significant degree in common disposal systems even though there are
many chemicals that can be successfully degraded when mixed with
an alkali or acid solution or in some cases a specially prepared
enzyme. The principal use would be in rinsing containers in
situations where the rinsate cannot be added to the mix. Data
requirements here would be dictated by the chemical involved and
the site of use as is the case in many other registration situations.

Developing label statements on disposal that are informa-
tive and fully supported by sound data will take time but the
effort is expected to be well worth while. Preparation of
guidelines is underway by EPA and input from any and all interested parties will be most welcome.

Literature Cited

1. Report of the Secretary's Commission on Pesticides and Their
    Relationship to Environmental Health, U.S. Department of
    Health Education and Welfare. December 1969.
2. U.S. Environmental Protection Agency. Pesticides and Pesti-
    cide Containers, Regulation for Acceptance and Recommended
    Procedures for Disposal and Storage. Federal Register,
    39(85):15236-15241, May 1, 1974.
3. Fox, A.S. and A.W. Delvo, 1972. "Pesticide Containers Assoc-
    iated With Crop Production," Proceedings of the National
    Conference on Pesticide Containers, New Orleans, November
    28, 1972, Published by the Federal Working Group on Pest
    Management, Washington, D.C.

4. Arthur D. Little, Inc. Economic Analysis of Pesticide Disposal Methods. Cambridge Massachusetts, Strategic Studies Unit, March 1977. EPA/540/9-77/018, 181 p.

5. Lawless, E.W., T.L. Ferguson, and A.F. Meiners. Guidelines for the Disposal of Small Quantities of Unused Pesticides. Midwest Research Institute, Kansas City, Missouri, June 1975. EPA/670/2-75/057, 342 p. (Available from the National Technical Information Service as PB-244 557)

6. Shih, C.C. and D.F. Dal Porto. Handbook of Pesticide Disposal by Common Chemical Methods. TRW Systems, Inc., Redondo Beach, California, December 1975. EPA/530/SW-112c, 109 p. (Available from the National Technical Information Service as PB-242 864)

7. Lande, S.S. Identification and Description of Chemical Deactivation/Detoxification Methods for the Safe Disposal of Selected Pesticides; Contract No. 68-01-4487, U.S. Environmental Protection Agency, Cincinnati, Ohio May 1978. 183p.

8. Day, H.R. Disposal of Dilute Pesticide Solutions. U.S. Environmental Protection Agency, Washington, D.C., Office of Solid Waste Management Programs, 1976. EPA/530/SW519, Day, 18 p. (Available from the National Technical Information Service as PB-261 160).

9. Wilkinson, R. R., E.W. Lawless, A.F. Meiners, T.L. Ferguson, G.L.Kelso and F.C. Hopkins. State of the Art Report on Pesticide Disposal Research. Midwest Research Institute, Kansas City, Missouri, September 1978. EPA/600/2-78-183, 225 p.

10. U.S. Environmental Protection Agency, Pesticide Registration: Proposed Data Requirements, Federal Register, Vol 47, No. 227, November 24, 1982.

11. U.S. Environmental Protection Agency, Pesticide Assessment Guidelines, Subdivision N, Chemistry, Environmental Fate, October 1982. EPA-540/9-82-021,

12. Sanborn, J.R., B.M. Francis, and R.L. Metcalf. The Degradation of Selected Pesticides in Soil: A Review of the Published Literature. Municipal Environmental Research Laboratory, Cincinnati, Ohio, EPA/600/9-77/022, 635 p. 1977 (Available from the National Technical Information Service as PB-272 353)

13. Ghassemi, M. and S. Quinlivan. A Study of Selected Landfills Designed As Pesticide Disposal Sites. EPA Publication No. SW-114c, 1975.

14. Ferguson, T.L., F.J. Berman, G.R. Cooper, R.T. Li, and F.I. Honea. Determination of Incinerator Operating Conditions Necessary for Safe Disposal of Pesticides. Midwest Research Institute, Kansas City, Missouri, December 1975. EPA/600/275 /041, 415 p. (Available from the National Technical Information Service as PB-251 131)

RECEIVED April 26, 1984

The Resource Conservation and Recovery Act

DAVID FRIEDMAN

Office of Solid Waste, Environmental Protection Agency, Washington, DC 20460

Background

RCRA mandates the EPA to identify those residuals which, if improperly managed, pose a hazard to either human health or the environment. In implementing the Act, EPA has promulgated regulations in Parts 260 through 270 of Title 40 of the Code of Federal Regulations. Parts 260 and 261 establish the conditions under which a material becomes a waste and identify those wastes which are "hazardous wastes" and must be managed according to the management standards of Parts 262 through 270.

Pesticide-containing materials may be classified as wastes subject to RCRA if they have served their intended use and are sometimes discarded, irrespective of whether they are being disposed of or are destined for recycling. A waste is a "hazardous waste" if it exhibits any of the characteristics of a hazardous waste, or is listed in sections 261.31, .32, or .33.

Pesticide wastes that are hazardous by reason of the characteristics are those which are either: solvent based and have a flash point $<60\ ^{\circ}C$; are aqueous and have a pH <2.0 or >12.5; release HCN or H_2S upon contact with acids; or leach greater than threshold levels of one or more of the elements arsenic, barium, cadmium, chromium, lead, mercury, selenium and silver, or the pesticides endrin, lindane, methoxychlor, toxaphene, 2,4-D or 2,4,5-TP. To date, these are the only pesticides for which thresholds have been established.

Listed hazardous wastes may be process residuals generated during the manufacture of the pesticide itself

(e.g., byproduct salts generated in the production
of cacodylic acid), materials resulting from the use
of the pesticide (e.g., bottom sediment sludge from
the treatment of wastewaters from wood preserving), or
discarded commercial pesticide products, off-specifica-
tion pesticides, container residues, and spill cleanup residues.

During 1981 approximately 25,835 metric tons of
listed pesticide wastes were generated in the United
States. Of these, 38% were pesticides manufacturing
residuals and the rest discarded commercial chemical
products. Of the 51 listed pesticides, only 19 were
reported to have been handled in quantity sufficient to
require RCRA control. It should be pointed out that
this survey represented 14,100 of the 60,000 firms who
requested ID numbers from EPA.

Generation/Transportation

Persons who generate more than 1000 kg/month of hazardous wastes are required to notify EPA and obtain a
generator ID number. In addition, the Agency annually
conducts a sampling of such generators to determine the
amounts of such waste generated and how it is disposed
of. Generators are also required to monitor the ship-
ment of their waste to insure that the waste is delivered to a treatment, storage or disposal facility that
is permitted to manage such wastes. In the event a
discrepancy exists between the amount of waste leaving
the generator's facility and the amount of waste reaching the TSD facility, the generator is required to
notify EPA of this within 45 days.

A farmer, however, disposing of waste pesticides
which are hazardous wastes, from his own use, is not
required to comply with the RCRA notification or
management standards provided he triple rinses each
emptied pesticide container and disposes of the pesticide residues on his own farm in a manner consistent
with the disposal instructions on the pesticide label.
This exemption from the RCRA management controls does
not apply, however, to commercial pesticide applicators.

Any person who transports, or offers for transportation, hazardous waste for off-site treatment,
storage, or disposal must prepare a manifest before
transporting the waste off-site. The person must designate, on the manifest, one facility which is permitted
to handle the waste described on the manifest. A person
may also designate one alternate facility that can be

used in the event an emergency prevents delivery of the waste to the primary facility.

The manifest must contain the name, address, telephone number and EPA identification number of the generator, transporter and designated receiving facility, and description, quantity, hazardous waste ID number, and hazard class of waste.

Management

All facilities engaged in the treatment, storage, or disposal of hazardous wastes must be permitted by either EPA or an authorized state. Two types of such permits exist. Those facilities which were in existence prior to November 1980, received an interim permit which allows continued operation, under the provision of Part 265, until final administrative disposition of the owner's or operator permit application is made.

All facilities operating under Interim Status are subject to a number of general facility standards. In addition, specific standards have been established for management of wastes in containers, tanks, surface impoundments, waste piles, landfarms, landfills, incinerators, and chemical, physical, and biological treatment facilities.

Facilities established after November, 1980 are required to obtain a permit before beginning operation. In order to obtain such a permit the owner/operator must demonstrate that the facility meets the standards established in 40 CFR Part 264.

Figure 1 is a Flow Diagram of the RCRA permitting process. As can be seen it includes a detailed technical evaluation of the design and proposed operating procedures of the facility to insure that no harm will occur to either persons or the environment from operation of the facility. The major steps in the process are:

-Applicant submits an application;
-EPA determines that the application is complete;
-EPA issues draft permit or notice of intent
    to deny permits;
-EPA gives public notice and solicits written
    comments; If deemed justified by the
    comments received, EPA holds a public hearing;
-EPA responds to comments and issues final
    decision.

            -Decisions, however, may be appealed to the
            EPA administrator or persons may seek
            judicial review.

        State agencies may receive authorization to
administer all or part of the hazardous waste program,
including issuance of permits. States are required to
have a program that is at least as stringent as the
Federal program. However, the programs do not have to
be identical. State requirements may differ in the
details of the permitting process and may be more
stringent.

        According to data EPA has recently received as
part of the reporting requirements of Part 262, during
1981, hazardous wastes regulated under RCRA were
managed in the following manner and amounts. As Figure
2 indicates, the method of management selected tends
to reflect the cost of such management.

Current Activities
Both Congress and EPA have a number of studies and
efforts underway which may ultimately impact disposal
of pesticide waste. Among these are: reevaluation of
the small generator exclusion limit, expansion of
Extraction Procedure toxicity characteristic to include
additional organic chemicals, revamping of Section 261.
33 (commercial chemical products which are hazardous
waste when discarded or intended to be discarded) to
both bring mixtures of active ingredients under the
definition and also to establish concentration thresholds for the wastes, and a prohibition on land disposal
of certain wastes. At this time I would like to briefly
touch on each of these areas.

        The RCRA regulations currently exempt persons who
generate or accumulate less than 1000kg per month of
most hazardous waste from regulatory control. The
Agency believed that this threshold exempts only about
1% of the total hazardous waste from RCRA control
while relieving 90% of the hazardous waste generators
of the regulatory controls. A study, however, is in
progress to reexamine the impact of this exemption.
If the Agency finds that significant amounts of hazardous waste are being mismanaged because of this exemption, changes may be proposed. A questionaire has
been prepared to survey small generators of hazardous
waste and we expect to send it out to approximately
50,000 generators by the end of the year. Congress,
however, is also examining the small generator exempt-

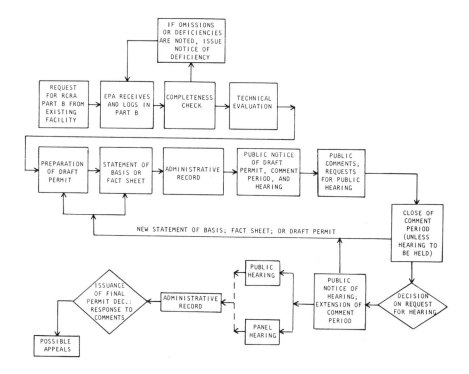

Figure 1. EPA's permitting process for existing land disposal facilities.

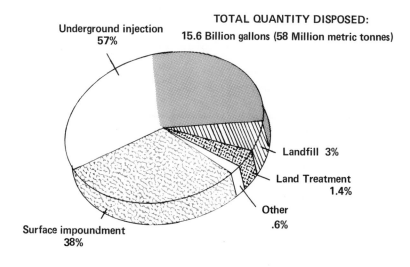

Figure 2. Quantities of hazardous waste disposed in 1981 (preliminary data).

ion as part of the RCRA reauthorization process and may legislate a lowering of the exemption threshold irrespective of Agency action.

Present hazardous waste identification characteristics generally do not include a means of identifying those wastes which pose a problem due to the presence of organic toxicants. The Extraction Procedure Toxicity characteristic only includes thresholds for six organic compounds. To correct this deficiency, the Agency is working to include additional toxic organic chemicals. The expansion may take the form that any waste containing any of the listed compounds at or above a certain threshold is a hazardous waste. The thresholds would be a compound specific and based on such factors as NOAELs (No Observed Adverse Effect Levels) and ADIs (Acceptable Daily Intake Levels). Any such expansion will likely include a number of pesticidal compounds.

Three other actions are being studied that may have a significant, if not substantial bearing on pesticide waste disposal under RCRA. These are (1) expansion of the listing of commercial chemical products that are hazardous waste when disposed of, (2) expansion of the discarded commercial chemical product listing to include products which are mixtures of active ingredients, and (3) establishment of concentration limits for the compounds on the list. While the first and second of these actions would act to bring a large number of commercial pesticide products under RCRA control, the third (establishment of threshold levels) would act to exempt many, if not most, application strength solutions from RCRA control. Once these thresholds have been established, products or solutions containing the toxic chemicals at concentrations below the threshold would automatically be excluded from regulation as a hazardous waste.

The final regulatory change being contemplated to be discussed is the land disposal ban. Both Congress and the Agency are concerned that for certain wastes land disposal may not be protective of human health and the environment for as long as the waste remains hazardous, taking into account the uncertainties associated with land disposal. In order to eliminate such practices, changes are being formulated to both the RCRA itself and the hazardous waste regulations which will prevent the land disposal of certain wastes. It is conceivable that some pesticidal wastes, especially those that are persistent, that have a potential to bioaccumulate, and are potentially leachable if placed in a non-secure landfill would be among those banned unless they have been treated by such methods as stabilization, neutralization, or destruction of toxic species.   In addition, the House and Senate are considering enactment of bans on disposal of both containerized and bulk liquids in landfills and a ban on underground injection of hazardous waste in Class IV wells.

In conclusion, the hazardous waste management regulatory system under RCRA is both dynamic and complex.   Pesticides, by their very nature, pose unique disposal problems.   It is thus incumbent upon persons managing such materials to keep abreast of current standards and regulations.   EPA has established a toll free "hotline" for such a purpose.   The telephone number is 800-424-9346 or if in Washington, D.C., 382-3000.

RECEIVED February 13, 1984

FIELD-DEMONSTRATION SCALE TECHNOLOGIES

Pesticide Waste Disposal in Agriculture

CHARLES V. HALL

Department of Horticulture, Iowa State University, Ames, IA 50011

Chemical compounds classified as pesticides (fungicides, herbicides and insecticides) total over
500. Others used as rodenticides, desiccants,
defoliants, etc., increase that number. Agricultural uses include all crops and livestock on
farms, plus those used as seed treatments, in
greenhouses, parks, golf courses, nurseries and
lawns in urban areas. Chemicals are applied by
farmers, industry, or institutional employees,
commercial applicators and individual citizens.
With proper planning the most common form of waste
to be disposed of is dilute rinse water.
Occasionally discontinued or non-usable concentrate
pesticides must and can be disposed of safely.
Greatest dangers of improper disposal are to water
supplies, food or feeds, recreation areas, animal
habitats, and other waste disposal facilities.
Three years research conducted jointly by 6 departments at Iowa State University and sponsored by the
U.S. Environmental Protection Agency has demonstrated that wastes from over 45 pesticides were safely
disposed of by containment in a concrete pit allowing evaporation of the liquid component, and
biodegradation, and other forms of pesticide decay.
Plastic lined pits were less satisfactory. A small
disposal pit suitable for individual farmers and
small applicators has been developed.

Pesticide Use

A prominent publisher (1) lists 678 compounds which are classed
as pesticides and used in the broad field of agriculture.
Included are insecticides, fungicides, herbicides, rodenticides,
growth regulators, etc. These compounds are often used to adjust
the biological balance in favor of the desired plant or animal

0097–6156/84/0259–0027$06.00/0

population being grown or the products being stored. They also
are used to completely eliminate the unwanted competitor. Many
are used to protect human health. The number of pesticides used
on a large scale in agriculture probably does not exceed 100 and
often they are very specific. Also, many are marketed in
different formulations. The relative volume of waste to be
disposed of is not proportional to the volume of pesticides used.
For example, the volume of herbicides used on corn, soybeans,
etc., is very large, but little waste is generated. Container
rinsate is recycled by triple rinsing back into the sprayer tank.
Commercial applicators use rinsates where the same chemical is
being used and only rinse the sprayer when a different pesticide
is being used or when discontinuing the operation.

Nature and Handling of Wastes

Often pesticide wastes, which require special disposal
facilities, are in a dilute form and result from rinsates from
containers, spray tanks, and equipment wash water. These may
originate from the small applicator or large commercial operator.
Such wastes should be sprayed on an area for which they are
approved or placed in a safe disposal facility. Occasionally,
fairly large volumes of recommended concentration dilute mixtures
resulting from livestock dipping operations, overestimating the
amount needed for a spray operation, etc., must be discarded.
For such operations, safe facilities or procedures are essential
to protect human health and environmental safety. If a hazardous
chemical, such as toxaphene is used, which requires many years to
degrade, the waste should be properly contained. However, most
organo-phosphates are readily biodegradable and can be spread on
land in accordance with label recommendations. In all cases,
disposal must be in accordance with the Federal Resource
Conservation and Recovery Act, and state and local regulations.
Pesticide wastes can and should be minimized by carefully
calculating the precise amount of pesticide needed and then
applying that entire amount on the area of intended use. All
liquid containers should be triple rinsed, punctured, and
disposed of in an authorized solid-waste facility or properly
recycled. Paper bags, plastic containers, etc., should be
properly incinerated or taken to an authorized solid-waste
facility where state and local regulations permit. In cases
where pesticides are discontinued, banned, flooded, out of date,
contaminated or fire damaged, it is necessary to dispose of
concentrated or formulated compounds. These are abnormal
situations and the state departments of environmental quality and
the U.S. Environmental Protection Agency officials provide
assistance in such emergencies. They should be notified
immediately as required by federal and state law. In many such
cases disposal can be accomplished over a period of time by
dilution, containment, biodegradation and evaporation.

Combustion may be the most satisfactory method for non-
biodegradable materials. The problem of disposal of long term
residual materials is of a lesser magnitude than 10 years ago due
to discontinued use, better planning, higher cost of chemicals,
and use of more rapidly biodegradable pesticides. In fact, some
pesticides currently in use biodegrade so readily that they are
limited in effectiveness as insecticides, herbicides, etc. (3).
However, it is important that non-biodegradable chemicals be
properly contained in accordance with federal regulation until
approved disposal can be accomplished.

A System for Safe Disposal of Pesticide Wastes

The disposal pit (Figures 1 and 2), used at the Horticulture
Station since 1970, was designed to contain surplus diluted
insecticides, fungicides, herbicides, growth regulators, etc.,
from spraying operations for fruit, vegetable, ornamental and
turfgrass research plantings. The farm consists of 229 acres
with diversified plantings. Therefore, the operation is typical
of many agricultural research and development centers located
throughout the U.S. in that a wide variety of different
pesticides are used which result in the generation of small
quantities of concentrate and larger amounts of diluted pesticide
mixtures. The system described in Table I was constructed to
provide a safe and satisfactory solution to the disposal of such
wastes. Waste from over 45 pesticides were disposed of in the
concrete pit between 1970-76 (Table II). Research was conducted
at Iowa State University by faculty in the Departments of
Agronomy, Agricultural Engineering, Energy and Mineral Resource
Institute, Entomology, Microbiology, and Horticulture. It was
sponsored by the U.S. Environmental Protection Agency over a
three year period to evaluate the effectiveness of current
disposal methods and develop new systems. In addition,
evaporation of dilute pesticide mixtures from a holding pit was
compared with water evaporation from a standard weather
evaporation pan and correlated with temperature, relative
humidity, sky conditions, wind direction, and velocity.
Evaporation models were developed for predicting evaporative
disposal needs for other geographic regions. Also, checks were
made for leakage and air pollution. All methods and models are
described fully in the final published report (2).
A new large pit was constructed at the Agronomy-Agricultural
Engineering Research Center with two thicknesses of 6-mil black
polyethylene plastic film as a liner. More intensive research
was conducted in 56 plastic minipits to evaluate chemical
interactions, degradation, and biological activity (2).
    Research results revealed that the concrete pit at the
Horticulture Station was safe from leakage, did not present a
hazard of air pollution, and allowed chemical and microbial
degradation of the deposited materials (2). The concrete pit, 12

Table I. Characteristics of the Disposal Pit shown in Figure 1.

Dimensions --   12 ft by 30 ft by 4 ft deep

Construction -- 8 inch reinforced concrete walls and bottom
                 with grooved connection and flexible ties.
                 (See (2) for details).
                 An automated mobile cover to allow for full
                 sun and wind exposure.
                 Has drain tile installed around base with
                 accessible riser for sampling for leakage.
                 Connected to the mixing room in adjacent
                 building by a pipe from the sump to permit
                 transfer of all wash water for evaporation.

Orientation --   on the west end of the pesticide and spray
                 equipment storage building with full south
                 and west exposure to sun and wind which
                 maximizes evaporation.
                 Raised above ground level to prevent
                 flooding by surface water from heavy rains.

Contents --   two one ft layers of coarse (3/4 - 1 1/2
                 in) washed river gravel with a one ft layer
                 of field soil containing in excess of three
                 percent organic matter in between (gravel-
                 soil-gravel).

ft by 30 ft by 4 ft deep, filled with a layer of gravel, one ft
of soil, and another layer of gravel, was effective for
evaporation of approximately 6000 gallons of liquid wastes
annually between Apr. 1 and Oct. 15 (Figures 1 and 2). The soil
layer within the pit contained relatively normal aerobic
bacterial activity during these months (4). The two primary
bacterial groups were Bacillus and Pseudomonas spp. No chemical
pollution was detected in the sampling tile located beneath the
pit, in the station well 50 yards away, or in the station lake
1000 yards down grade from the disposal site. The system is
effective at present after 13 seasons of use.
     Pesticide containers were triple rinsed, crushed and
disposed of as solid waste (Figure 3). Containment of liquid
wastes by the newly constructed plastic lined pit was
questionable after one year. There appeared to be some leakage
or fluctuation of the liquid level. There is continual danger of
rupture of such liners by mechanical injury, chemical
interaction, rodents, etc., which could result in contamination

Figure 1. The concrete disposal pit with automated mobile cover and adjacent pesticide storage facility.

Figure 2. Same as Figure 1 with cover closed.

Table II. Pesticides used at the Horticulture Station 1970-76.
Small amounts of leftover diluted materials were
deposited.

Compound	Compound
Alachlor	Guthion
Atrazine	Heptachlor
Azinphos methyl	Hexachlorobenzene
Benomyl	Kelthane
Bensulide	Lannate
Butralin	Malathion
Captan	Mancozeb
Carbaryl	Maneb
Chlorothalonil	MCPP
Chloroxuron	Methomyl
Citcop	Methoxychlor
2,4-D	Metribuzin
2,4-DB	Naptalam
DCPA (Dacthal)	Omite
Diathane M-22, M-45, and Z-78	Paraquat dichloride
Dicamba	Penoxalin
Dichlobenil	Phosmet
Diphenamid	Polyram
Endosulfan I and II	Propachlor
EPTC (Eptam)	Simazine
Ethylparation	Sulphur
Folpet	Trifluralin
Glyphosate	

of subsurface water where the water table is high (5).
Certainly, two 6 mil polyethylene layers would be inadequate for
long term containment, especially if equipment is to be driven
over the fill surface. In more arid regions the problem would be
of lesser magnitude for most commonly used agricultural
pesticides and especially where the water table is 200-300 feet
deep and there is a deep clay subsoil layer between. However,
local regulations must be considered in each case to ensure
environmental safety.

Summary and Current Status

Based on research sponsored by the U.S. Environmental Protection
Agency and long term experience at Iowa State University some
essential components of safe disposal of agricultural pesticide
wastes were: 1) dilution, 2) containment in a structure that will

Figure 3. Hydraulically operated can crusher with five gallon cans before and after crushing.

not leak, overflow, flood or otherwise pollute the environment,
3) evaporation of the water, and 4) biodegration of most
compounds.

The system in use is too large and elaborate for most farm,
greenhouse, nursery, golf course, or small park operations. A
new precast concrete micropit was installed at the Iowa State
University Horticulture Station in 1983 which may serve as a
model for such individual operators (Figures 4 and 5). The same
functional components used in the macropit were incorporated to
provide maximum evaporation, biodegration, and environmental
safety. Previous attempts to use plastic, fiberglass and other
containers were unsuccessful because of freezing, thawing, and
rupturing problems in winter. This structure should withstand
those conditions and incorporates the gravel-soil-gravel system
previously used. The cover is similar to that suggested for
modification of the large pit and pipes are installed to permit
sampling for leakage. Multiple units could easily be installed
depending on evaporative needs and local evaporation rates. The
same precautions should be used to avoid flooding and to maximize
evaporation. Also, similar units should be available from local
concrete products companies throughout the country.

The structure (12 ft by 30 ft by 4 ft deep) has been in use
at the Horticultural Station since 1970 and during the three
years intensive research, was used to dispose of over 6000
gallons of liquid each year or the equivalent of approximately 35
surface inches per year. No contamination of surrounding soil,
water, or air was detected. Therefore, the system was found to
be environmentally safe, however, some modifications could be
made which would improve overall efficiency of use and retain
effectiveness (Table III).

Acknowledgments

Funding for research conducted and reported (2) was provided
under U.S. Environmental Protection Agency Grant No. R804533,
Cincinnati, Ohio.

Other faculty actively involved in the research were: James
Baker-Agricultural Engineering, Paul Dahm-Entomology, Loras
Freiburger-Horticulture, Layne Johnson-Microbiology, Gregor Junk-
Energy and Mineral Resources Institute, Fred Williams-
Microbiology and Charles J. Rogers-U.S.E.P.A. as Project Officer.
Original macropit design was by Thamon Hazen and revisions by
Dennis Jones-Agricultural Engineers.

Journal Paper No. J-11207 of the Iowa Agriculture and Home
Economics Experiment Station, Ames, Iowa. Project No. 2216.

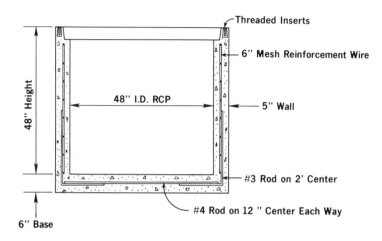

Figure 4. Structural specifications for a modified precast manhole structure as revised and redrawn from Iowa Concrete Products Co. SK-83-61.

Figure 5. The above concrete unit as installed at the I.S.U. Horticulture Station as a small quantity disposal unit.

Table III. Suggested Modifications for the Concrete Macropit.

1. Install a raised fixed cover of opaque corregated fiberglass in hinged sections which slope to prevailing sun and with sufficient overhang to prevent rain from entering. The cover should be designed to withstand maximum wind velocities for region where located.

2. Enclose the disposal pit with 1/2 in mesh hail screen attached to cover support posts, to keep children and animals from entering and debris from collecting on the pit surface.

3. Install an enclosed wash rack for equipment in an adjacent structure with drain connected through a sump (pump installed) for disposal of wash water from tank and equipment. Wash rack must have a cleanout trap for removal of soil and other debris from equipment.

4. Install a recirculating pump with a mist system for enhancement of evaporation in more humid climates.

5. Design capacity to needs based on environmental conditions of the region and state, and local regulations.

6. Provide adequate sampling tubes, or tiles, to conform to federal and state monitoring regulations.

Literature Cited

1. Farm Chemicals Handbook 82. Meister Pub. Co. p. C 3-318.
2. Hall, Charles V. et al. "Safe Disposal Methods for Agricultural Pesticide Wastes". National Technical Information Service. May 1981. PB-81-197 584.
3. Fox, Jeffery L. Soil Microbes Pose Problems for Pesticides. Science. Vol. 221, No. 4615, pp. 1029-30. Sept. 9, 1983.
4. Johnson, Layne M. and Paul A. Hartman. Microbiology of a Pesticide Disposal Pit. Bull. of Environm. Contom. Toxical. 25, 448-455 (1980).
5. Haxo, H. E., Jr. Interaction of Selected Lining Materials with Various Hazardous Wastes. II. Proceedings of the Sixth Annual Research Suymposium. U.S.E.P.A. pp. 160-180. March 1980.

RECEIVED March 6, 1984

Degradation of Pesticides in Controlled Water–Soil Systems

G. A. JUNK, J. J. RICHARD, and P. A. DAHM[1]

Ames Laboratory, Iowa State University, Ames, IA 50011

Atrazine, alachlor, 2,4-D ester, trifluralin, carbaryl, and parathion were added individually and as mixtures to 60 L of water and 15 kg of soil held in 110 L plastic garbage containers that were buried partially in open ground. Degradations from initial pesticide concentrations of 0.4 and 0.02 weight percent were investigated. Additional variables of aeration at 1 L/min and peptone nutrients at 0.1 weight percent, as possible aids to degradation, were also studied. Aliquots from 56 buried containers were taken for chemical analyses at regular intervals during a 68 week period. These samples were analyzed for the added pesticides and their hydrolysis products. Conclusions based on analytical results for the field experiment and supplementary laboratory experiments are: 1) soil and water in an inexpensive container provide for satisfactory containment of common pesticides so that chemical and biological degradations can occur; 2) soil is essential for containment and is a satisfactory source of microorganisms; 3) aeration and addition of buffers, nutrients and inoculants are of questionable value; 4) the half-life concept for degradation is not applicable; 5) sampling of disposal sites, even small controlled ones, is a problem; 6) degradations vary from rapid for hydrolysis of 2,4-D ester and carbaryl to unobservable for atrazine.

[1]Current address: Department of Entomology, Iowa State University, Ames, IA 50011

0097-6156/84/0259-0037$08.75/0
© 1984 American Chemical Society

Studies were initiated at Iowa State University in 1977 to determine if pesticides would be contained and degraded when deposited
in water/soil systems. Although the addition of known amounts of
the selected pesticides was controlled, the physical environment
was not; temperature, humidity, wind speed, etc. were normal for
the climate of Central Iowa. Four herbicides and two insecticides
were chosen on the basis of three factors. Firstly, they represented six different families of pesticides. The four herbicides,
alachlor, atrazine, trifluralin, and 2,4-D ester, represent the
acetanilides, triazines, dinitroanilines, and phenoxy acid herbicides, respectively. The two insecticides, carbaryl and parathion, represent the carbamate and organophosphorus insecticides,
respectively. Secondly, the pesticides were chosen on the basis
of current and projected use in Iowa (1) and the Midwest. Thirdly, the chosen pesticides were ones for which analytical methodology was available.
        Considerable information has been published concerning the
degradation of these pesticides. A brief summary of their degradation pathways and their expected persistence in the environment
is presented here.

Atrazine

The major degradation of atrazine in soil was its conversion to
hydroxyatrazine by loss of the chlorine atom (2-5). Dealkylation
also occurred with deethylation predominating over deisopropyla-
tion (5,6). Only small amounts of the radioactivity of the ring
labeled atrazine was converted to $^{14}CO_2$ by soil (6,8-10). Geller
(11) found that the percentages of $^{14}CO_2$ evolved from ^{14}C-labeled
side chains were similar for biological and nonbiological dealkylation. No distinction could be made between the two processes so
degradation was assumed to be initiated by abiotic environmental
factors, such as low pH, mineral salts, organic matter and photolysis. The s-triazines are one of the most recalcitrant groups
of herbicides and persistence of over one year in the soil was observed when atrazine was applied at the recommended rates (12).

Alachlor

Most acetanilides are biodegraded rapidly in soil, but alachlor
appears to be degraded by a mechanism different from that for
other members of this group of herbicides. The presence of either
the 2´,6´-dialkyl substituents, the N-alkoxylmethyl substituent,
or both, may preclude enzymatic hydrolysis of the carbonyl or

amide linkages of alachlor (13). Hargrove and Merkle (14) reported that 2-chloro-2´,6´-diethylanilide was formed in alachlor-
treated, air-dried soil incubated at 46°C. This degradation product was shown to result from acid catalyzed hydrolysis on mineral
surfaces. Beestman and Deming (15) found a half-life of 7.8 days
for alachlor in unsterilized soil. The average persistence for
recommended rates of application was 6-10 weeks (12).

Trifluralin

Both aerobic and anaerobic degradation pathways have been proposed
for trifluralin (16). Dealkylation is the initial aerobic degradation followed by sequential removal of the second alkyl group to
give the dealkylated product. Reduction of the two nitro groups
eventually leads to the formation of the 3,4,5-triamino-α,α,α-
trifluorotoluene. Under anaerobic conditions the nitro groups are
reduced first, followed by dealkylation, with formation of the
same 3,4,5-triamino-α,α,α-trifluorotoluene product. Degradation
of trifluralin was more rapid and extensive in substrate-amended
soil under anaerobic conditions compared with well-aerated systems. The relative rates followed the order, moist anaerobic >
flooded anaerobic > moist aerobic (17). Degradation in these
environments after 20 days was 99, 45 and 15%, respectively.
Under field conditions trifluralin has been predicted to degrade
to nonphytotoxic levels within a growing season when soil conditions are moist and warm (12). After three years, less than 1.5%
of ^{14}C-trifluralin was detected in test plots maintained under
natural conditions (18).

2,4-D Ester

Evidence for the microbiological degradation of 2,4-D ester in
soils was based on the stimulation by warm, moist conditions and
organic matter (19); a correlation between degradation rate and
the numbers of aerobic soil bacteria (20); and inhibition when the
soils were air-dried and autoclaved (19). Little information is
available, however, on the nature of the degradation products.
The results of many studies of the degradation of phenoxyalkano-
ates using pure cultures of microorganisms have been reported.
The use of Arthrobacter isolated by Loos (21) has been studied
especially well . This organism was isolated from silt loam and
it rapidly oxidized 2,4-D. The first intermediate was 2,4-dichlo-
rophenol. Eventually the aromatic ring of the 2,4-D was cleaved
with all the bound chlorine converted to free chlorine.

The results of Smith (22) suggested that the isopropyl and n-butyl esters of 2,4-D are subject to rapid chemical hydrolysis in soils; however the isooctyl ester was more stable and possibly undergoes some biological hydrolysis. Bailey et al. (23) reported that the hydrolysis of the propylene glycol butyl ester in pond water was 90% complete in 16-24 hours and 99% complete in 33-49 hours. Zepp et al. (24) reported that the half-life for hydrolysis of various 2,4-D esters varied from 0.6 hour for the 2-butoxyethyl ester to 37 hours for the 2-octyl ester. Persistence of the ester in the soil environment was estimated to be less than one week. The degradation of the 2,4-D acid was also rapid (12) but slower than the hydrolysis of the ester.

Parathion

Degradation of parathion in soil was by hydrolysis to p-nitrophenol and diethylthiophosphoric acid and reduction to aminoparathion (25,26). Chemical oxidation of parathion in soils and waters was not prevalent, although oxidation of the phosphorus-sulfur bond has been shown to occur under ultraviolet light and in oxidizing environments (26). At ordinary levels of application to soil, parathion was degraded within weeks if microbial activity was available (27). Accumulations even after repeated applications were unlikely (28). When higher concentrations were applied to soil, persistence increased. Simulated spills of concentrated parathion resulted in a 15% residue after five years (29) and 0.1% after 16 years (30).

Carbaryl

Carbaryl degradation was primarily microbiological as reported by a number of investigators (31-36). Soil organisms transformed carbaryl to many metabolites, including 1-naphthol, 1-naphthyl N-hydroxy methyl carbamate, 1-naphthyl carbamate, and 4 and 5-hydroxy-1-naphthyl methyl carbamate. The degradation of 1-naphthol also occurred microbiologically (32,35) by a pathway similar to hydroxylation with subsequent ring cleavage of naphthalene (37). Predicted persistence of carbaryl in the environment varied from one to several weeks (38). The degradation of 1-naphthol was predicted to be faster (34,35) than degradation of carbaryl.

EXPERIMENTAL

Description of System

Atrazine (Aatrex; 80W), alachlor (Lasso; E.C.) 2,4-D ester (Weedone LV4; E.C.), trifluralin (Treflan; E.C.), carbaryl (Sevin; 50W) and parathion (Security; 15W) were blended, individually and as mixtures, with 60 L of water and 15 kg of sandy loam soil in 110 L plastic garbage containers buried partially in the ground. A cross-sectional view of one of these buried containers is shown in Figure 1. The individual pesticides were studied in separate containers at the high and low concentrations of 0.4 and 0.02 weight percent active ingredient. The mixtures were studied with all six pesticides each present at high and low concentrations in separate containers. Additional variables of aeration at 1 L/min and peptone nutrients at 0.1% by weight resulted in a factorial experiment of 56 containers. The layout of these containers in a 7x8 matrix, with dots showing the locations, is shown in Figure 2. The high and low concentrations are indicated by 1X and 0.05X, respectively. Those systems under aeration and with nutrients are also identified. All 56 containers, for studying the degradations of these six pesticides, and the associated equipment were laid-out in a fenced area covering only 60 M^2.

Sampling

A 100 g sample of the soil and liquid contents of each container was taken for analyses at 1, 3, 4, 8, 12, 16, 20, 24, 28, 52 and 68 weeks after addition. The samples were obtained by slowly lowering and raising a 100 mL bottle, capped with a two-hole rubber stopper, through the swirl formed by vigorous mixing of the contents of the container. The mixing was accomplished using a propeller blade attached to a shaft driven by a variable speed drill. The amount of water necessary to adjust the volume to the original 60 L was recorded before mixing and sample collection.

Extraction Procedures

The collected sediment and water samples were centrifuged for 0.5 hour at 200 rpm. The water was decanted and the volume measured prior to transfer to a 250 mL separatory funnel where it was extracted four times with four 50 mL volumes of diethyl ether for the high concentration samples and with 50 mL followed by two 25 mL volumes for the low concentration samples. The 2,4-D samples only were acidified to a pH of ~2 with H_2SO_4 to aid in the solvent extraction.

For the four herbicides, the soil fraction in the sample bottle was extracted by adding 50 mL of diethyl ether followed by agitation for 15 minutes on a wrist-action shaker. The diethyl ether was decanted and the high concentration samples were extracted three more times with 50 mL of diethyl ether by hand shaking the capped bottles for 2 to 3 minutes. The low concentration samples were extracted two additional times with 25 mL of diethyl ether.

For the two insecticides, the soil was extracted with 75, 50 and 50 mL of an acetone:benzene:methanol (1:2:1 by vol.) mixture. The sample bottles were capped and agitated for 60 minutes on a wrist-action shaker for each extraction. The contents of the bot-

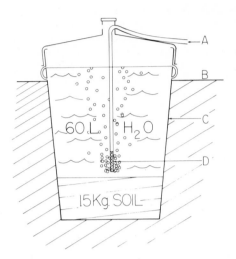

Figure 1. Cross-sectional view of a buried garbage can. A,D--
Tygon tube and diffuser used for aeration; B--ground level;
C--110L plastic garbage container.

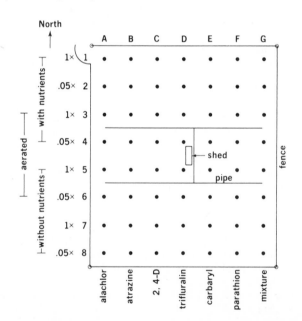

Figure 2. Layout of the 7 x 8 or 56 matrix of garbage cans
containing 6 pesticides at two different concentrations and
mixtures with and without aeration and nutrients.

tles were then centrifuged for 15 minutes and the supernatant
liquid was decanted. The combined liquid from the three extractions was reduced to about 25 mL under partial vacuum. The liquid
was extracted in a separatory funnel with three 50 mL portions of
diethyl ether. The combined diethyl ether extracts were filtered
through anhydrous Na_2SO_4 and the filtrate reduced to 50 mL under
partial vacuum.

Separation and Analyses

Atrazine, alachlor and trifluralin were determined in the extracts
of the samples by gas chromatography using a N-P detector. An EC
detector was used for the determination of parathion, 2,4-D ester
and the 2,4-D acid after esterification with diazomethane. Carbaryl and 1-naphthol were determined colorimetrically by the procedure of McDermott and DuVall (39).
      Recoveries for soil and water samples spiked individually
with the six commercial formulations at the 0.4 and 0.02% levels
were 96 and 97%, respectively. Comparable recovery efficiencies
were obtained for mixtures of the pesticides. In addition, the
ability to account for all the deposited pesticides in the
analyses of the samples taken from the containers prior to any
degradations was validation for the effectiveness of the
extraction procedures.

RESULTS AND DISCUSSION

Soil and Liquid Analyses

The analytical data for the added pesticides and two of the
hydrolysis products, 2,4-D acid and 1-naphthol, were used to
formulate the degradation graphs shown in Figures 3-14. Atrazine
underwent no degradation either alone or in mixtures and alachlor
and trifluralin underwent no degradation in mixtures, so the
graphs for these pesticides under these conditions are not shown.
      For the sake of clarity, some of the analytical data from 1,
3, and 4 weeks have been averaged and plotted as a single result
at the four week interval. These graphs indicate vividly which
pesticides degrade and what factors such as concentration, aeration, mixtures, and nutrients affect the rate of degradation. The
graphs also indicate the inevitable uncertainty in the analytical
results, due to errors in collecting samples from a heterogeneous
medium.
      The placement of the graphs show the effect of concentration,
where several pesticides decay readily at low levels but do not
show measurable degradation when present at high concentration.
The rate of degradation for an individual pesticide when it is
alone or in mixtures can be compared by inspecting successive
figures. For example, Figure 3 shows the plots for 2,4-D ester
when it is present alone at two different concentrations and under

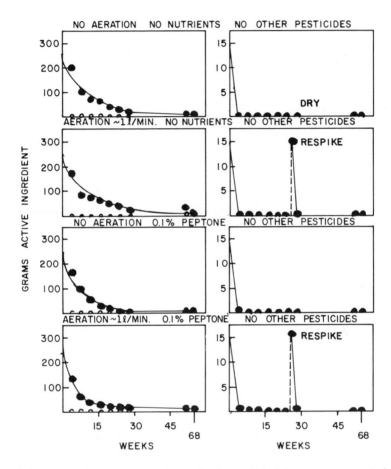

Figure 3.   2,4-D ester degradation with time. •, amount in soil and water; o, amount in water.

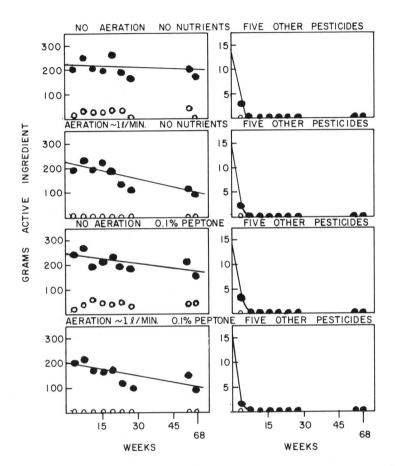

Figure 4. 2,4-D ester degradation with time in presence of five other formulated pesticides. ●, amount in soil; o, amount in water.

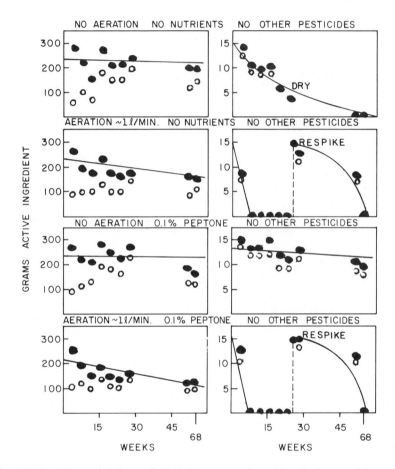

Figure 5.   Degradation of 2,4-D ester plus the 2,4-D acid hydrolysis decomposition product with time. ●, amount in soil and water; o, amount in water.

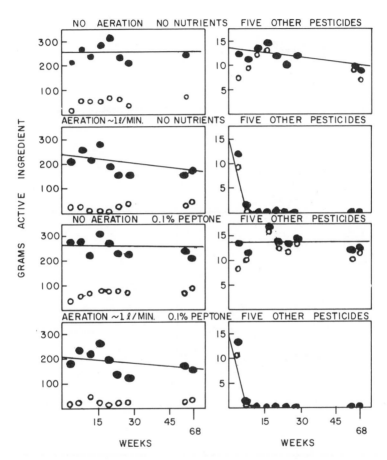

Figure 6.   Degradation of 2,4-D ester plus the 2,4-D acid hydrolysis decomposition product with time in the presence of five other formulated pesticides. ●, amount in soil and water; o, amount in water.

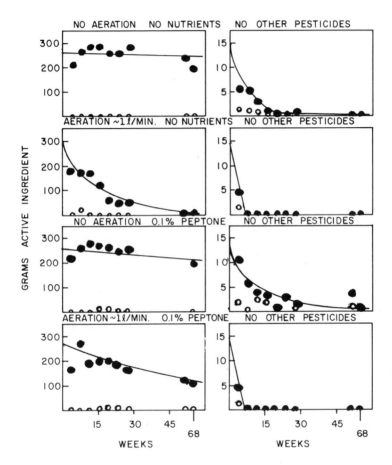

Figure 7. Carbaryl degradation with time. •, amount in soil and water; o, amount in water.

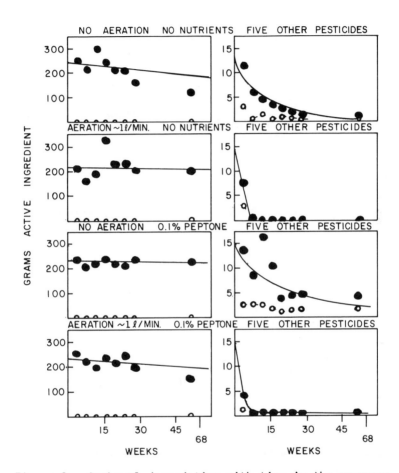

Figure 8. Carbaryl degradation with time in the presence of five other formulated pesticides. ●, amount in soil and water; o, amount in water.

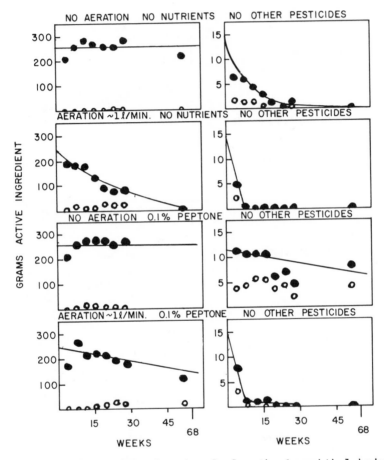

Figure 9. Degradation of carbaryl plus the 1-naphthol hydrolysis decomposition product with time. ●, amount in soil and water; o, amount in water.

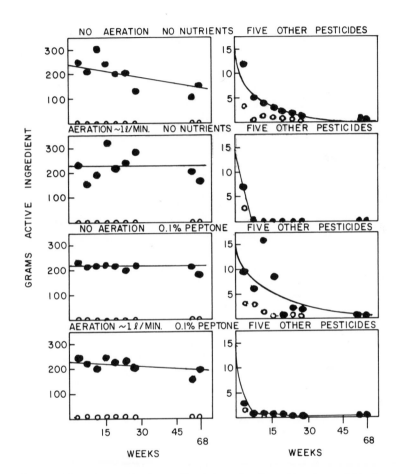

Figure 10.   Degradation of carbaryl plus the 1-naphthol hydrolysis decomposition product with time in the presence of five other formulated pesticides. ●, amount in soil and water; o, amount in water.

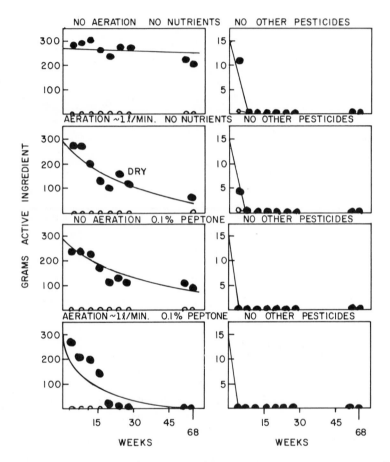

Figure 11. Parathion degradation with time. ●, amount in soil and water; o, amount in water.

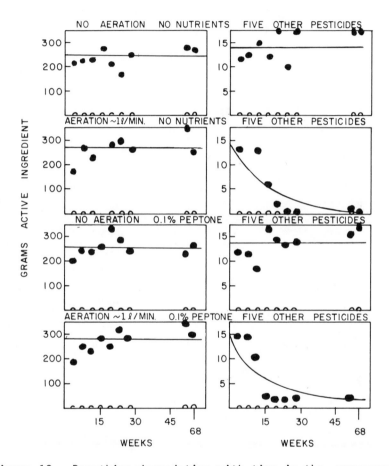

Figure 12. Parathion degradation with time in the presence of five other formulated pesticides. ●, amount in soil and water; o, amount in water.

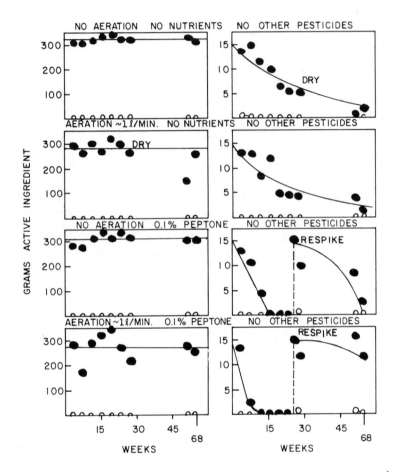

Figure 13. Trifluralin degradation with time. •, amount in soil and water; o, amount in water.

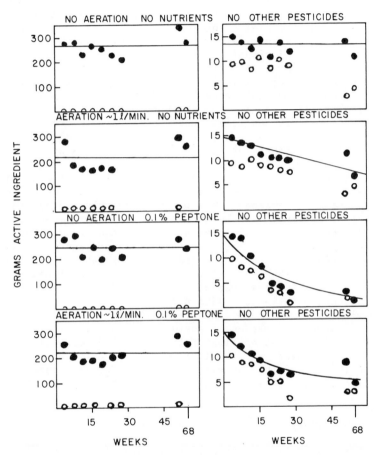

Figure 14. Alachlor degradation with time. ●, amount in soil and water; o, amount in water.

different conditions of aeration and nutrients. In Figure 4 these
same conditions are repeated for 2,4-D ester present with comparable amounts of each of the other five pesticides. The presence of
the other pesticides slows down the degradation of 2,4-D ester.
This general inhibitory effect of mixtures can be seen for the
other pesticides by comparing Figures 5 and 6 and so on through
Figures 11 and 12. These figures along with the data for degradation of trifluralin and alachlor in Figures 13 and 14 are also
very useful for highlighting the most favorable conditions for
degradation of an individual pesticide but they do not provide a
good visual representation of the degradation of the pesticides
relative to each other. This is best accomplished by referring to
plots like those shown in Figures 15 and 16. Figure 15 shows the
degradation at low concentration corresponding to 15 g of active
ingredient of the six pesticides and two hydrolysis products under
ambient conditions of no aeration, no nutrients and no mixtures.
The same representations for larger amounts of 300 g are given in
Figure 16.

The plots also can be used to obtain detailed information
about the degradation of the different pesticides at any one point
in time. These comparisons, when weighed properly against the
analytical uncertainty, can be used to make deductions about the
effects of the variables included as part of the study. For example, Figure 11 shows that the addition of peptone nutrient has a
more pronounced effect on the degradation of parathion than does
aeration. However 15 g amounts of parathion are completely degraded within eight weeks even at ambient conditions in the absence of either aeration or added nutrients. Many other deductions similar to this example can be made from inspection of these
figures. However, summation of these deductions results in a
rather confusing representation of the overall effects of aeration, nutrients, mixtures, concentration and chemical structure on
the effectiveness of the container systems. The influence of
these variables can be ascertained best from the degradation summary given in following section.

Degradation Summary

The total analytical data, some of which are plotted in Figures 3-
14, were used to formulate the matrix summary shown in Table I.
For a generalized interpretation of the matrix, the six pesticides
chosen for the study are assumed to be representative of recalcitrant to easily degradable pesticides. The criteria for a positive (YES) degradation was reduction to less than 10% of the
amount originally deposited into the containers within the reasonable interval of 68 weeks. The reader should bear in mind that
only about 30 of these 68 weeks were conducive to degradation for
the seasonal climatic conditions where these containers were
located in Iowa. A more quantitative matrix, which includes factors for the amount of degradation at all time periods would not

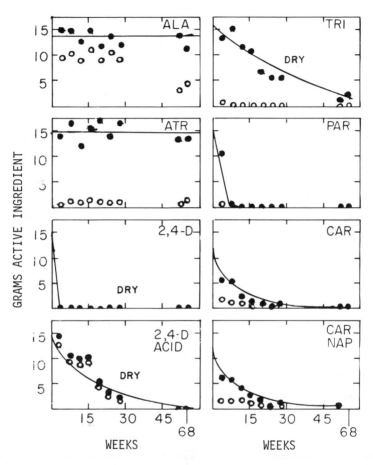

Figure 15. Degradation of alachlor (ALA), trifluralin (TRI), parathion (PAR), 2,4-D ester (2,4-D), carbaryl (CAR), 2,4-D acid (ACID) and 1-naphthol (NAP) at low concentration and ambient conditions. ●, amount in soil and water; o, amount in water.

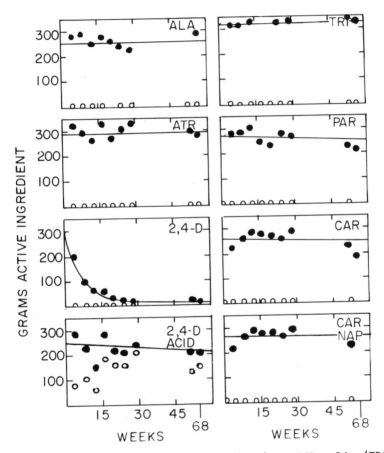

Figure 16. Degradation of alachlor (ALA), trifluralin (TRI), parathion (PAR), 2,4-D ester (2,4-D), carbaryl (CAR), 2,4-D acid (ACID) and 1-naphthol (NAP) at high concentration and ambient conditions. ●, amount in soil and water; o, amount in water.

Table I   Summary of the Effect of Ambient Conditions, Aeration at 1 L/min, Peptone Nutrient at 0.1 Wt. %, Concentration, and the Presence of 5 Other Pesticides (Mix) on the Degradation of Six Pesticides and Two Decomposition Products

Condition	Initial amount (g)	Alachlor Alone	Alachlor Mix	Atrazine Alone	Atrazine Mix	2,4-D Ester Alone	2,4-D Ester Mix	2,4-D Acid Alone	2,4-D Acid Mix	Trifluralin Alone	Trifluralin Mix	Parathion Alone	Parathion Mix	Carbaryl Alone	Carbaryl Mix	1-naphthol Alone	1-naphthol Mix	Yes count†
								Positive degradation to <10% active ingredient within 68 weeks*										
No nutrients: No aeration	15	No	No	No	No	Yes	Yes	Yes	No	Yes?	No	Yes	No	Yes	Yes	Yes	Yes	9
	300	No	No	No	No	Yes	No	No	No	No	No	No	No	No	No	No	No	1
Aeration	15	No	No	No	No	Yes	Yes	Yes	Yes	Yes?	No	Yes	Yes	Yes	Yes	Yes	Yes	11
	300	No	No	No	No	Yes	No	No	No	No	No	Yes	No	Yes	No	Yes	No	3
Nutrients: Aeration	15	No	No	No	No	Yes	Yes	Yes	Yes	Yes	No	Yes	No	Yes	Yes	Yes	Yes	10
	300	Yes?	No	No	No	Yes	No	No	No	No	No	No	No	Yes?	No	Yes	No	2
No aeration	15	No	No	No	No	Yes	Yes	Yes	No	Yes	No	Yes	No	Yes	Yes	Yes	Yes	8
	300	No	No	No	No	Yes	No	No	No	No	No	No	No	No	No	No	No	1
Yes count#		1	0	0	0	8	4	3	2	4	0	5	1	5	4	5	3	45

*Only 32 of which were conducive to degradation because of drastic changes in seasonal temperature. Any borderline cases based on known analytical deviations are given a question mark following the Yes for degradation to <10% of the initial amount. All systems were active microbiologically over the entire 68 week period.

#A Yes count of 8 indicate very easy degradability — a count of 0 indicate strong persistence. Mixture effect shown in column pairs.

†A Yes count of 16 would indicate complete degradability of all components at listed conditions — concentration effect shown in row pairs.

be very useful for interpretation purposes because of the possible bias related to the analytical uncertainties. These more subtle comparisons are best made by close inspection of the individual degradation graphs with due regard for the uncertainty in the analytical results.

The matrix summary provides a basis for decisions regarding the addition of nutrients, the incorporation of aeration mechanisms, the change in the disposal load, the restriction on the dump of certain pesticides, etc. Some observations about degradations from inspection of the matrix summary are presented in the following paragraphs.

The effects of all combinations of aeration and nutrients are easily discernable on the right hand YES count column. The increase from 10 to 14 degradation units on a scale of 0 to 32 when aeration was incorporated must be weighed carefully against the cost of adding aeration to the containers. Nutrients had an inhibitory, if any, effect on pesticide degradation in general.

Concentration had a dramatic effect on degradation but the actual amount degraded per unit time is unknown and would require future tests after longer periods of warm weather. For the present, the effect of a 20X concentration increase (15 to 300 g) on a degradation scale of 0 to 64 was reflected by the 38 degradation units for low concentration (15 g) compared to only 7 for high concentration (300 g).

Mixtures had an inhibitory effect with the most dramatic being the degradation of 2,4-D ester, trifluralin and parathion. On a scale of 0 to 64, the individual pesticide degradations decreased from 31 to 14 when mixtures were present. This apparent antagonistic action of mixtures was probably more of a latency effect related to the six-fold increase in total pesticide concentrations rather than a toxic action to the microorganisms.

For degradations in separate containers and a scale of 0 to 8, atrazine (0) was the most persistent pesticide followed in order by alachlor (1), 2,4-D acid (3), trifluralin (4), carbaryl (5), 1-naphthol (5), parathion (5), and 2,4-D ester (8).

This degradation order also existed when mixtures and individual pesticides were considered together on a scale of 0 to 16 degradation units.

For degradation when mixtures were present and a scale of 0 to 8, atrazine (0), alachlor (0), and trifluralin (0) were all persistent, followed in order by parathion (1), 2,4-D acid (2), 1-naphthol (3), carbaryl (4), and 2,4-D ester (4).

Other less important observations can be made by the reader by breaking down the relative degradations into pairs of low and high concentrations and using scales of 0 to 4 degradation units.

Additional observations and conclusions based on interpretation of the degradation data are as follows:
1. The effect of temperature on the rate of degradation can be seen by inspecting the degradation plots in Figures 5 and 13 where Weedone and Treflan had been respiked. The respiking

occurred at the 28th week during the late autumn. These two
herbicides had originally degraded very rapidly during the
warm temperatures of spring and early summer. The cold temperatures of late autumn and winter resulted in inhibition of
the degradation until warmer temperatures occurred again in
the following spring sometime near the 60th week.

2. Only an insignificant loss of pesticides occurred when the
liquid contents of the containers were lost due to ruptures
caused by the freeze-thaw cycles during the winter months.
Almost complete containment was achieved by transferring the
solid contents to a new plastic container and adding fresh
water. The word DRY on the figures indicate those systems
where these losses of liquid occurred.

3. The water data plotted in Figures 3-16 showed that most of the
pesticides are sorbed on soil particles. This confirmed what
is probably the most beneficial effect of soil in the containers; it guaranteed almost complete containment even when the
liquid contents were lost.

4. Buffers were not helpful for the degradation of large amounts
of 2,4-D ester.

5. No measurable losses due to purging of the pesticides occurred
during aeration. The surprising lack of a purging effect was
attributed to the presence of soil and the closed lid system.

Water Loss, pH and Temperature

On each of the days when samples were taken for analyses, the
evaporation losses of water and the pH of the container contents
were measured. Intermittently the temperatures of the container
contents were measured.

The evaporative losses of H_2O in liters are plotted for 1977
in Figure 17. Similar results were obtained for 1978. In general, the greatest losses occurred during the hot summer months with
total losses per year averaging about 10 L. In continuously hot
climates with temperatures of ~25°C, the water evaporation losses
from similar closed lid systems would be about 2 L/month or 24
L/year. This represents the maximum volume of liquid waste which
a container can handle per month and per year. The 28 systems
that were aerated did not show any greater H_2O losses than the
other 28 systems. This surprising result would probably not be
obtained if an open lid design were used in a continuously hot
region.

The pH measurements of the mixed contents were made immediately after taking the soil and liquid samples. The presence of a
very large amount of suspended sediment affected the accuracy and
reproducibility of these pH measurements. The average values for
each container are summarized in Table II. The deposited pesticides, the formulating agents and the decomposition products did
not change the pH from that observed for ambient water except for

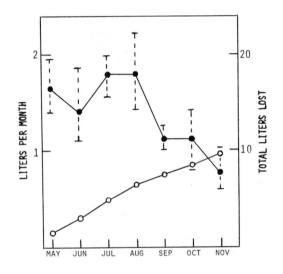

Figure 17. Average H_2O loss per month (●) and average total
H_2O loss (o) during 1977 for 56 systems. (Brackets indicated
the maximum and minimum loss each month.)

those systems having high concentrations of 2,4-D ester. The analytical data for these containers showed that rapid hydrolysis to the acid occurred. It was this acid that caused the low pH. However even this pH was not too low for survival of most all microorganisms. Indeed the counts of microorganisms from these particular systems were no different than for any of the other systems (40). The lowering of the pH did not occur when Weedone was present with the five other pesticide formulations. In some cases this was due to slower degradation when a mixture was present but even when the analytical data showed that the ester had been hydrolyzed, no pH near 5 was observed. Obviously, the substances used in the other five formulations acted as a buffer.

All of our data suggested that pH of the liquid contents was of concern only if large amounts of acidic pesticides, or esters which are easily hydrolyzed, are being dumped. This is not to say that pH is unimportant for some degradations. The pH may well be a consideration for a particular pesticide but it seems to be unimportant for a general disposal system. Nevertheless the pH should be checked regularly because it can be done so easily. If necessary then, buffers can be added if the pH becomes low enough to affect the microorganisms or the degradation of some chemicals of concern.

Table II. pH Summary

Pesticide	Ave pH at Conc	
	High	Low
Alachlor	6.8	6.8
Atrazine	6.9	6.8
2,4-D ester	4.7	6.8
Trifluralin	6.7	6.7
Carbaryl	7.2	7.0
Parathion	6.7	6.9
Mixture	6.4	6.9

The temperature of the contents of the containers was not measured regularly. In general the ground acts as a damper for short term and seasonal changes in temperature. For example, the containers still had ice in them in April long after the ground had thawed. On the other end of the temperature scale, the container contents were still liquid in November long after stagnant surface waters had frozen. In mid-summer, the temperatures of the container contents were stable at about 22°C.

Laboratory Degradation Studies

Several bench-scale degradation experiments were completed. The data from these studies of 2,4-D butoxyethanol ester, alachlor, trifluralin and atrazine under a variety of aerobic and anaerobic conditions were used for the following preliminary observations: 1) ambient soil was about as good an inoculant as were the aerobic and anaerobic slimes and sludges; 2) artificial nutrients were not very useful for degradation purposes; 3) continuous mixing was of questionable value; 4) photolytic effects were not measurable for degradation of atrazine which was recalcitrant under all conditions.

Additional experiments with high concentrations of 2,4-D butoxyethanol ester with and without buffers showed no difference in degradation rates which confirmed the observations made in field experiments. Based on interpretation of results of these laboratory studies, combinations of buffers, anaerobic sludge, aerobic slime, light and constant mixing were not considered to be necessary for small and intermediate sized disposal systems.

Evidence of Degradation

In the experimental design used for these studies, the evidence of degradation was usually indirect in that only the deposited pesticide was monitored. When the amounts decreased with time, chemical and biological degradation mechanisms were assumed because our evaporation and leakage studies showed that the losses could not be accounted for by these routes. However, some losses by routes which could not be detected by the analytical regime remained a possibility. Thus, the best evidence of degradation was the observation of primary products. If the concentration of a primary degradation product increased concurrent with a decrease in concentration of the parent pesticide, then this direct evidence was even more convincing. However, the increase in the primary degradation products did not always occur due to further rapid degradations to less complicated components.

Those degradation products which have been identified in our investigations are 1-naphthol from carbaryl, 2,4-D acid and 2,4-dichlorophenol from 2,4-D ester, 2-chloro-2´,6´-diethylacetanilide from alachlor, α,α,α-trifluro-2-nitro-6-amino-N,N-dipropyl-p-toluidine and α,α,α-trifluro-2,6-diamino-N,N-dipropyl-p-toluidine from trifluralin, and a variety of phenols and acids from the degradation of the aromatic solvents used in the formulation of the liquid pesticides as emulsifiable concentrates (41,42).

CONCLUSIONS

1. Inexpensive plastic garbage cans are satisfactory containment units for the disposal of small amounts of formulated pesticides.

2. Soil and water are essential to an effective disposal system. The soil is a source of microorganisms, a food source and a great aid to containment because of its high adsorption capacity. The water is an aid in monitoring and for microbiological activity.
3. Addition of aeration and nutrients is of questionable value in disposal systems especially if installation cost and maintenance are considered. The use of other possible methods of enhancing degradation such as aerobic slime and anaerobic sludge are also questionable.
4. Volatilization losses should be insignificant since even the very volatile trifluralin was not lost.
5. Atrazine and alachlor are both very persistent.
6. The use of half-life to estimate the persistence of a pesticide in the natural environment is a very debatable practice. If used, one should be cautious about the probable errors due to different ambient conditions and the unpredictable concentration effect. This same caution applies to the extrapolation of other laboratory results to field situations.
7. Collecting valid samples, after mixing these small containers holding only soil and liquids, is a problem. Large systems containing soil, liquids and rocks that can't be mixed present an even more severe sampling problem as evidenced by the discussion of segregation given in Chapter 5.

ACKNOWLEDGMENTS

Part of the work was performed in the laboratories of the U.S. Department of Energy and supported under contract No. W-7405-Eng-82. This work was partially supported by the Office of Health and Environmental Research, Office of Energy Research. The assistance of E. Catus, C. V. Hall, R. S. Hansen, T. E. Hazen and H. J. Svec in the administration of the EPA contract No. R804533010, (C. Rogers, project officer) which provided the major support of this research, is appreciated. The authors especially acknowledge the work of James Baker in coordinating and setting up the disposal site and Joyce Chow, Greg Gorder, Grace Schuler and Fred Williams for assistance in the sampling and analyses.

LITERATURE CITED

1. Jennings, V.; Stockdale, H. Coop. Ext. Serv. Pub. 1978, PM-845, Iowa State Univ.; Ames, Iowa.
2. Harris, C. I. J. Agr. Food Chem. 1967, 15, 157-162.
3. Knuesli, E. D.; Berrer, O.; Dupuis, G.; Esser, H. in "Herbicides: Chemistry, Degradation and Mode of Action"; Kearney, P. C.; Kaufman, D. D., Eds.; Marcel Dekker, Inc.: N. Y., 1975; Chap. 2.
4. Roeth, F. W.; Lavy, T. L.; and Burnside, O. C. Weed Sci. 1969, 17, 202-205.

5.   Beynon, K. I.; Stoydin, G.; Wright, A. N. Pestic. Biochem.
     Physiol. 1972, 2, 153-161.
6.   Skipper, H. D.; Volk, V. V. Weed Sci. 1972, 20, 344-347.
7.   Sirons, G. J.; Frank, R.; Sawyer, T. J. Agr. Food Chem. 1973,
     21, 1016-1020.
8.   Skipper, H. D.; Gilmour, C. M.; Furtick, W. R. Proc. Soil
     Sci. Soc. Amer. 1967, 31, 653-656.
9.   McCormick, L. L.; Hiltbold, A. E. Weeds 1966, 14, 77-82.
10. Goswami, K. P.; Green, R. E. Environ. Sci. Technol. 1971, 5,
     426-429.
11. Geller, A. Arch. Environ. Contam. Toxiol. 1980, 9, 289-305.
12. "Herbicide Handbook of the Weed Science Society of America",
     Champaign, Ill., 5th ed. 1983.
13. Kaufman, D. D. in "Pesticides in Soil and Water"; Guenzi, W.
     D., Ed.; Soil Science Society of America, Inc.: Madison, Wis.
     1974; Chap. 8.
14. Hargrove, R. S.; Merkle, M. G. Weed Sci. 1971, 19, 652-654.
15. Beestman, G. B.; Deming, J. M. Agron. J. 1974, 66, 308-311.
16. Probst, G. W.; Golab, T.; Wright, W. L. in "Herbicides:
     Chemistry, Degradation, and Mode of Action"; Kearney, P. C.;
     Kaufman, D. D. Eds.; Marcel Dekker, Inc.: N.Y., 1975; Vol. 1,
     Chap. 9.
17. Parr, J. E.; Smith, S. Soil Sci. 1973, 115, 55-63.
18. Golab, T.; Althaus, W. A.; Wooten, H. L. J. Agric. Food Chem.
     1979, 27, 163-179.
19. Hernandez, T. P.; Warren, G. F. Proc. Am. Soc. Hort. Sci.
     1950, 56, 287-293.
20. Akaminc, E. K. Bot. Gaz., 1951, 112, 321-319.
21. Loos, M. A. in "Herbicides: Chemistry, Degradation and Mode
     of Action"; Kearney, P. C.; Kaufman, D. D., Eds. Marcel
     Dekker, Inc.: N.Y. 1975; Vol. 1, Chap. 1.
22. Smith, A. E. Weed Res. 1972, 12, 364-372.
23. Bailey, G. W.; Thurston, A. D.; Pope, J. D.; Cochrane, D. R.
     Weed Sci. 1970, 18, 413-419.
24. Zepp, R. G.; Wolfe, N. L.; Baughman, G. L.; Gordon, J. A.
     "Environmental Quality and Safety"; Coulston, F.; Korte, F.
     Eds. Georg Thieme: Stuttgart, 1975; pp. 313-317.
25. Graetz, D. A.; Chesters, G.; Daniel, T. C.; Newland, L. W.;
     Lee, G. B. J. Water Pollut. Control Fed. 1970, 42, 76-94.
26. Lichtenstein, E. P.; Schulz, K. R. J. Econ. Entomol. 1964,
     57, 618-627.
27. Lichtenstein, E. P.; Fuhremann, T. W.; Schulz, K. R. J. Agr.
     Food Chem. 1968, 16, 870-873.
28. Goldsworthy, M. C.; Foster, A. C. Am. Fruit Grower 1950, 70,
     No. 2, 22; 52-53.
29. Wolfe, H. R.; Staiff, D. C.; Armstrong, J. F.; Comer, J. W.
     Bull. Environ. Contam. Toxicol. 1973, 10, 1-9.
30. Stewart, D. K.; Chisholm, D.; Ragab, M. T. H. Nature 1971,
     229, 47.

31. Johnson, D. P.; Stansbury, H. A. J. Agr. Food Chem. 1965, 13, 133–138.
32. Kazano, H.; Kearney, P. C.; Kaufman, D. D. J. Agr. Food Chem. 1972, 20, 975–979.
33. Rodriguez, L. D.; Wyman Dorough, H. Arch. Environ. Contam. Toxicol. 1977, 6, 47–56.
34. Sikka, H. C.; Miyazaki, S.; Lynch, R. S. Bull. Environ. Contam. Toxicol. 1975, 13, 666–672.
35. Bollag, J. M.; Liu, S. Y. Can. J. Microbiol. 1972, 18, 1113–1117.
36. Bollag, J. M.; Liu, S. Y. Soil Biol. Biochem. 1971, 3, 337–345.
37. Davies, J. I.; Evans, W. C. Biochem. J. 1964, 91, 251–261.
38. Mount, M. E.; Oehme, F. W. Carbaryl: A literature review Residue Rev. 1981, 80, 1–64.
39. McDermott, W. H.; Duvall, A. H. J. Assoc. Off. Anal. Chem., 1970, 53, 896–898.
40. Hall, C. J.; Baker, J.; Dahm, P.; Freiburger, L.; Gorder, G.; Johnson, L.; Junk, G.; Williams, F. Safe Disposal Methods for Agricultural Pesticide Wastes, U.S.E.P.A. Report 600/2-81-074, NTIS Accession PB81 197584 (1981).
41. Junk, G. A.; Richard, J. J. Water Qual. Bull. 1981, 6, 40–42.
42. Junk, G. A.; Richard, J. J. in "Advances in the Identification and Analysis of Organic Pollutants in Water"; Keith, L. H., Ed.; Ann Arbor Science: Ann Arbor, Mich., 1981; Chap. 19.

RECEIVED February 13, 1984

Pesticide Disposal Sites: Sampling and Analyses

G. A. JUNK and J. J. RICHARD

Ames Laboratory, Iowa State University, Ames, IA 50011

Pesticides and their degradation products were analyzed in samples taken from two disposal pits located at Iowa State University, Ames, IA. The first
was an eight-year-old 30,000 L concrete-lined pit
where over 50 kg of more than 40 different pesticides had been deposited. The second was a two-
year-old 90,000 L polyethylene-lined pit where 150
kg of 24 different pesticides had been deposited.
The pesticide concentrations in the soil and liquid
samples taken from these pits showed extreme variations which necessitated collecting and compositing
samples from many different points to estimate the
average pesticide concentrations and their change
with time. Water, soil and air samples were also
collected and analyzed to evaluate the possible
contamination of the surrounding environment. Summarized conclusions from these investigations are:
1) pit disposal systems are effective in containing
many pesticides to the extent that release to surrounding air and water is insignificant; 2) possible
environmental effects can be established by avoiding
the extreme difficulties associated with solid samples and taking only liquid samples for analyses;
3) pit disposal systems are effective for pesticides
as well as other organic chemicals.

0097-6156/84/0259-0069$07.75/0

The three main goals of the disposal of dilute solutions of waste
pesticides are containment, detoxification and volume reduction.
To meet these goals two pits have been constructed for the disposal of pesticide waste connected with horticulture and agronomy
operations at Iowa State University. This paper deals with the
sampling and chemical analyses of the soil and liquid contents
from these pits plus water, soil and air samples from the surrounding area to help answer questions concerning containment and
degradation of the deposited pesticides.

EXPERIMENTAL

Description of Disposal Pits

Detailed descriptions of the two disposal pits located at the
Horticulture and Agronomy farms at Iowa State University, Ames,
IA, have already been published (1). Brief summaries of the con-
structions and operations are presented below for those who do not
have easy access to that report (1).

Horticulture Pit. The horticulture pit was constructed in 1969-70
and has been in continuous use since that time. The 8.8 x 3.4 m
concrete pit has an average depth of 1 m and has been filled with
alternate 30 cm layers of gravel and soil as shown in Figure 1.
The soil is a silt loam (Clarion-Nicolett-Webster) characteristic
of Central Iowa. A cover closes automatically to prevent precipitation from entering. A tile system constructed below the concrete floor of the pit connects to a sump for sampling of the
ground water. The 30,000 L pit has had over 50 kg of more than 40
different pesticides deposited since beginning operation.

Agronomy Pit. The pit at the Agronomy-Agricultural Engineering
Research Center was constructed in 1977 and used for the 1978 and
1979 growing seasons. As shown in Figure 2, the pit is 1.7 m deep
with surface dimensions of 13.7 x 7.3 m. The total volume of the
pit is ~90,000 L. Five vertical tile, 15 cm in diameter, are used
for liquid sampling. A galvanized metal pipe in an H shape is
buried within the rock and used as a distribution system. A nip-
ple extends above the rock surface for coupling to outlets of
sprayer rigs. The waste pesticide mixtures are thus distributed
over the pit area by gravity flow and drainage from the holes

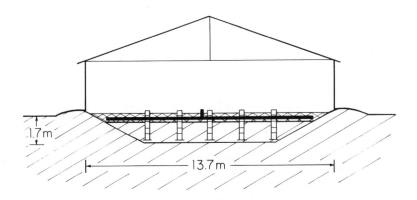

Figure 1. Cross-section of covered 8.8 x 3.4 m concrete-lined Horticulture pit with alternate layers of rock (cross-hatched) and soil (slanted lines).

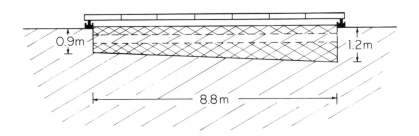

Figure 2. Cross-section of canopied 13.7 x 7.3 m plastic-lined Agronomy pit with a distribution system (solid bar) embedded in rock (cross-hatched) and five sampling tiles extending through the rock and soil (slanted lines).

drilled in the pipe. The wooden canopy prevents precipitation
from entering the pit. The pit was lined with two sheets of continuous black polyethylene encompassed on both sides by layers of
sand. Metallic hardware cloth was also laid on the outside soil-
sand interface to stop burrowing rodents. The pit was backfilled
to a depth of 1.22 m with the original silt loam soil, and then a
0.46 m layer of crushed rock was added to the surface. Since
beginning operation in 1978, ~150 kg of 24 different pesticides
have been deposited.

Sampling

Horticulture Pit. Core samples of soil of about 100 g size to a
depth of 10 cm and liquid samples of 500 mL were taken from eight
sampling points spaced uniformly across the pit surface. Each
sample was then solvent extracted individually and the pesticides
and degradation products were separated and quantitated.

To establish whether a composited sample would give results
equivalent to the average of the individual analyses, eight samples were taken in October 1977. A portion of each sample was
formed into a single composite and all samples were then analyzed.
Results for the average of eight individual samples and the composite agreed within 2%. Therefore all samples after October 1977
were composited.

Water samples were also collected from the drainage tile
surrounding this concrete-lined pit. Other water samples were
taken from sites remote from the pit location. Air samples for
pesticide vapor analyses were collected in the vicinity of the
pit.

Agronomy Pit. Liquid samples only were collected from five access
tiles at the Agronomy pit. These were analyzed individually and
as a composite, with results agreeing within 2%.

No soil samples were taken from within the Agronomy pit. One
set of soil samples to a depth of 3 m was taken from 10 cm diameter holes augered at each outside corner of the pit. Water samples from outside the pit were also collected from the seepage of
ground water into the augered holes and from other sites remote
from the pit location. Air samples for pesticide vapor analyses
were collected in the vicinity of the pit.

Extraction Procedures

The procedures employed for the liquid, soil, water and air samples are described and discussed below.

Liquid Samples. Three procedures were evaluated for extracting
pesticides from liquid samples collected at the Horticulture and
Agronomy pits. They were the resin sorption method of Junk et al.
(2), solvent extraction with hexane-diethylether, and solvent

extraction with methylene chloride. Comparable recoveries were achieved for all three procedures. The methylene chloride procedure was chosen because of the ready availability of pure solvent and the simplicity of the method.

Liquid samples collected at the Agronomy pit were partially stabilized suspensions which required pretreatment prior to the extraction. The pretreatments which were investigated were 24 hour settle, 24 hour settle plus one hour centrifuging at 2500 rpm, and 24 hour settle plus one hour centrifuging plus filtering through a 1 to 2 micron filter. These three methods were used for the liquid samples collected in August of 1978. The results are tabulated in Table I. For most of the pesticides comparable results were obtained with the amounts after centrifuging and filtering generally being lower. These samples also had the lowest turbidity and thus were more representative of true dissolved constituents. All samples collected after August 1978 were settled for 24 hours and then centrifuged and filtered before extraction.

Table I. Concentration Difference in Liquid Samples from Agronomy Pit when Settling, Centrifuging and Filtration are Employed

Herbicide	PPM		
	24 hour Settle	Centri-fuge	Centrifuge + Filter
EPTC	7	4	2
Butylate	13	7	3
Chlorpropham	15	8	6
Propachlor	132	110	32
Atrazine	279	193	174
Trifluralin	189	108	51
Alachlor	318	223	171
Cyanazine	97	119	109
Metribuzin	111	100	138

Soil Samples. Five methods of extraction were evaluated for soil samples collected at the Horticulture pit. The results for this evaluation are tabulated in Table II. The Woolson procedure (3) was selected for use throughout the course of our investigations of all soil and sediment samples.

Water Samples. The pesticide residues in ground water samples from sites adjacent to and remote from the disposal pit locations and the well water samples were analyzed by the resin sorption method using XAD-2 as described by Richard et al. (4) and Junk et al. (5) for measuring very low amounts of selected pesticides.

Table II. Comparison of Various Solvents and Extraction Proce-
           dures for Recovery of Pesticides from a Horticulture Pit
           Sample

Pesticide	PPM by Method Number[†]				
	1	2	3	4	5
Bensulide	26	16	22	24	24
Chlorothalonil	162	184	193	196	146
DCPA	1116	1141	1129	1054	1017
Dichlobenil	4	2	3	5	4
Endosulfan I	38	37	36	36	35
Endosulfan II	23	23	22	22	21
Hexachlorobenzene	89	91	89	82	79
Methoxychlor	93	95	93	93	91
Phosmet	4	2	3	3	2
Trifluralin	36	30	22	20	22

[†]1 = batch extraction using hexane-acetone according to procedure
     by Woolson (1974)
  2 = ultrasonic extraction using isopropyl alcohol
  3 = ultrasonic using hexane-acetone
  4 = ultrasonic using benzene-isopropyl alcohol
  5 = soxhlet extraction using hexane-acetone

Air Samples. The atmospheres in the vicinity of the pits were
sampled for pesticide vapors using XAD-2 resin and a vacuum pump
(6). Collection efficiencies for this method were measured in the
laboratory using simulated atmospheres and found to be 99.5% average for 10 pesticides spiked at 10 ng/L of air.

Separation and Detection Procedures

The majority of the pesticides were separated by gas chromatography and detected by a variety of general and specific detectors
such as flame ionization, electron capture, flame photometric,
electron impact and chemical ionization mass spectrometric and
nitrogen-phosphorus. Two columns were most useful for separating
the pesticides in the extracts of the samples collected from the
Horticulture and Agronomy disposal pits. These two columns were a
4% SE-30/6% OV210 and a 10% DC-200. A 30 m SE-54 glass capillary
column was employed for more complete separations when necessary
to aid in identification and analyses procedures. Relative retention times of several pesticides and methyl esters are given in
Table III for the SE-54 glass capillary column.

Some of the pesticides deposited in the disposal pits were
not gas chromatographable directly. These required special treatment to give products that could be determined by gas chromatography or by other procedures. These pesticides and the procedures employed for their analyses are discussed below.

Benomyl was extracted and isolated by the procedure of Austin
and Briggs (7). After isolation and concentration, benomyl and
its metabolites were determined colorimetrically.

Mancozeb, Maneb, Polyram and Dithane in soil and water samples were analyzed using the head space procedure for screening
food samples for dithiocarbamate pesticide residues as described
by McLeod and McCully (8).

Paraquat was determined in soil and water samples by the
extraction and gas chromatographic procedure of King (9).

Acidic pesticides and metabolites were concentrated from
aqueous solution by the anion procedure of Richard and Fritz (10).
The anionic materials in these concentrates were methylated using
diazomethane and the derivatized products were separated and detected by gas chromatography. Test results of the recovery efficiencies by this method for several pesticides and suspected metabolites have been reported elsewhere (11). An overall recovery
of 93% was achieved for sixteen acidic pesticides and metabolites
spiked into water at 200 ppb.

RESULTS AND DISCUSSION

Segregation

Individual analyses of both soil and liquid samples taken from the
Horticulture pit established unequivocally the extreme segregation
of the pesticides. Variations in concentrations for samples collected at eight sampling points in October 1977 were extreme as
shown in Table IV, being ~1100 for DCPA in soil samples and ~50
for phosmet in water samples. Similar evidences of segregation
were also found for soil and water samples collected earlier in
March, June and August of 1977. The greatest difference in concentration for soil samples was observed in June where DCPA varied
from 4 to 13400 ppm. For water samples, the maximum segregation
was for hexachlorobenzene, also in June, which varied from 0.1 to
50 ppm. These wide ranges in concentrations are probably due to
uneven depositions of the waste pesticides and their very slow
mobility in systems containing rock, soil and a liquid phase.

Similar segregation was observed in liquid samples collected
from the Agronomy pit where a system was installed to uniformly
distribute the pesticides. Concentrations of pesticides varied
by a factor as high as 10 in the June 1978 liquid samples as shown
in Table V.

The extreme variation in the chemical composition of samples
taken from different locations within the Horticulture and Agron-

Table III. Relative Retention Times of Various Pesticides or
    Their Methyl Esters (M.E.) on a SE-54 Capillary Column

Pesticide	t_R	Pesticide	t_R
Trimethylphosphate	0.09	Dimethoate	0.85
Dimethylthiophosphate	0.18	Nitroamino treflan	0.87
Dimethyldithiophosphate	0.22	Dinitramine	0.88
2,4-Dichlorophenol-M.E.[†]	0.33	Fluchloralin	0.88
Linuron	0.34	2,4,5-T-M.E.	0.89
Dichlobenil	0.42	Chlorothalonil	0.92
Eptam	0.43	Bentazon-M.E.	0.93
2,6-Dichlorobenzoic Acid-M.E.	0.45	Metribuzin	0.95
Tetrahydrophthalimide-M.E.	0.49	Endosulfan I	0.96
Butylate	0.50	Alachlor	0.97
2,4,5-Trichlorophenol-M.E.	0.51	Chloroxuron	0.98
Tetrahydrophthalimide	0.59	Aldrin	1.00
Dicamba-M.E.	0.63	Metolachlor	1.02
MCPA-M.E.	0.63	Malathion	1.03
MCPP-M.E.	0.64	DCPA	1.04
Propachlor	0.71	Ethyl Parathion	1.05
Chlorpropham	0.74	Heptachlor Epoxide	1.07
2,4-D-M.E.	0.74	Cyanazine	1.08
BHC	0.74, 0.80	Butralin	1.09
Trifluralin	0.78	Penoxalin	1.09
Pentachlorophenol-M.E.	0.79	Captan	1.11
Hexachlorobenzene	0.79	Folpet	1.11
Lindane (γ)	0.81	Kelthane	1.15, 1.21
Dealkylalachlor	0.81	p,p´-DDE	1.20
Chloramben-M.E.	0.82	Dieldrin	1.22
Benzimidazole	0.83	Endosulfan II	1.22
Atrazine	0.84	p,p´-DDD	1.27
Diamino treflan	0.85	p,p´-DDT	1.33
Simazine	0.85	Methoxychlor	1.42
Profluralin	0.85	Naptalam-M.E.	1.47

Temperature programmed at 5°/min from 80-270°
[†]M.E. = methyl ester

Table IV. Pesticide Concentrations at the Eight Sampling Points in the Horticulture Pit – October 1977

Pesticide	Sample Type	PPM at Sampling Point Number							
		1	2	3	4	5	6	7	8
Azinphos-Methyl	H$_2$O	0.3	5.5	0.9	4.4	1.1	4.5	0.6	0.2
	Soil	384	94	157	2902	765	84	396	24
Chlorothalonil	H$_2$O	0.2	1.2	0.2	1.1	<0.1	0.7	1.0	1.5
	Soil	108	10328	25	287	298	29	77	177
DCPA	H$_2$O	0.9	1.0	0.5	1.7	0.8	0.7	1.2	0.5
	Soil	492	639	8.5	9738	984	72	1026	38
Dichlobenil	H$_2$O	<0.1	<0.1	<0.1	<0.1	0.1	<0.1	0.1	0.1
	Soil	1.2	1.1	0.9	2.4	1.9	0.4	4.2	0.3
Endosulfan I	H$_2$O	0.4	0.3	0.2	0.8	0.4	0.4	0.3	0.2
	Soil	42	13	4.2	60	21	1.2	3.8	1.9
Endosulfan II	H$_2$O	0.2	0.2	<0.1	0.5	0.3	0.2	0.2	<0.1
	Soil	18	9.8	2.0	37	9.8	1.9	3.4	<0.1
EPTC	H$_2$O	12	7.6	1.2	19	10	7.2	19	9.4
	Soil	8.1	6.2	3.3	11	5.1	3.2	1.7	<0.1
Hexachlorobenzene	H$_2$O	<0.1	<0.1	<0.1	0.2	<0.1	<0.1	0.2	<0.1
	Soil	50	85	0.7	77	74	8.6	124	6.5
Phosmet	H$_2$O	<0.1	2.0	<0.1	5.4	1.6	2.8	<0.1	<0.1
	Soil	<0.1	<0.1	<0.1	1.6	8.6	1.7	<0.1	<0.1
Propachlor	H$_2$O	9.2	7.1	4.4	36	15	1.8	28	4.3
	Soil	<0.1	<0.1	<0.1	<0.1	<0.1	<0.1	<0.1	<0.1

Table V. Herbicide Concentrations at the Five Sampling Points in
          the Agronomy Pit - June 1978

Pesticide	PPM at Sampling Point Number				
	1	2	3	4	5
Alachlor	1646	4681	819	1347	1667
Atrazine	54	146	22	50	10
Butylate	200	130	24	81	125
Chlorpropham	110	210	25	193	314
Cyanazine	220	843	110	235	284
Dimethoate	<1	4	3	<1	4
EPTC	44	38	24	40	50
Propachlor	226	442	144	306	369
Trifluralin	1395	370	989	2372	3256

omy pits makes it difficult to collect a valid sample representative of the pit contents. Therefore an accurate mass balance is
nearly impossible to achieve. At best a rough estimate can be
obtained if sufficient samples are taken for analyses. These
results show the folly of attempting to estimate the amount of
pesticides or other hazardous chemicals in disposal pits, landfills, dump sites and even sediments when only a limited number of
areal and depth samples are taken.

Horticulture Pit

The accumulations of the amounts of pesticides deposited in the
Horticulture pit are given in Table VI for the period June 6, 1977
to October 1, 1979. Even though a very conscientious effort was
made to record all depositions, this pit is referred to as an
uncontrolled system since the amounts and the timing of the dumps
were at the discretion of the farm operator rather than the analytical chemist. For example, an appreciable amount of Polyram was
deposited prior to June 6, 1977 and none was dumped from June
through October 10. Sometime after October 1977, but before May
1978, an additional amount was dumped to bring the total to 2392
g. However the late growing seasons of 1978 and 1979 were different than 1977 because some further dumps of Polyram were made
during the summer and fall. Thus the analytical data for the
various pesticides will be a perturbation of the total amount deposited, the timing of the dumps and the sampling, the chemical
degradation, the biological degradation, the volatilization and
losses due to incomplete containment. Under this kind of operational and analytical mode, the best opportunity to document the
operation of the disposal system is to review the data for formulated pesticides such as Dacthal, Ramrod, Captan, Imidan, Polyram,

Table VI. Disposal Accumulations at the Horticulture Pit in Grams of Active Ingredient (AI)

Formulation[†] Deposited	Grams AI at Listed Dates					
	Jun 6 1977	Oct 10 1977	May 31 1978	Nov 3 1978	Jun 1 1979	Oct 1 1979
Polyram[*]	1974	1974	2392	2699	3108	3244
Imidan[*]	730	1651	1718	2319	2535	2716
Captan[*]	1050	2050	2126	2616	2678	3338
Thiodan[*]	42	753	753	1009	1009	1027
Paraquat[*]	29	132	184	249	470	522
Ramrod[*]	1335	2273	2567	4495	4495	5435
Bravo[*]	6	619	619	1752	1752	2275
Marlate[*]	39	108	147	2242	2328	2427
Tenoran[*]	569	815	815	1381	1381	1381
Sevin[*]	2	402	402	991	991	1200
Dymid	229	1367	1367	3113	3113	3113
Treflan	138	138	160	160	569	569
Benlate[*]	12	14	35	271	302	337
Guthion[*]	12	14	14	30	30	30
Prowl	4	4	4	4	4	4
Lasso	4	4	4	4	4	4
Amex	13	13	13	13	13	13
Dacthal[*a]	3379	7926	8011	10340	10381	12167
Simazine	129	129	160	160	414	414
Eptam	975	975	975	1127	1127	1127
Lannate[*]		57	57	221	221	221
Dithane M-45		274	274	735	735	735
Malathion[*]		245	245	591	591	633
Kelthane[*]		386	386	386	386	387
Roundup			44	44	44	44
Casoron[*]			12	12	295	295
Alanap			2325	2325	2325	2325
Prefar			1237	1237	1237	1237
Folpet			136	240	240	240
Butoxone				22	22	35
Dithane M-22				154	154	154
2,4-D[*]				369	369	369
MCPP				185	185	185
Dicamba				37	37	37
Aatrex 80W					37	37
Sencor						27
TOTAL +	10671	22323	27182	36793	43579	48301

Unknown amounts of these formulations and Amiben, Carbaryl, Chlordane, Cythion, Heptachlor, Maneb, Pyrethrum, Toxaphene and Zineb were deposited prior to the commencement of record keeping beginning June 1977. In addition to the listed pesticides 2619 g of Citcop, NAA, N-Serve, Dipel, Omite and Metasytox were deposited but not analyzed.

[a]Hexachlorobenzene is an ~15% impurity.

etc. which have been deposited in large amounts continuously over
a three year period. The absence of any appreciable build-up is
then an indicator of system utility.

Soil and Liquid Analyses. The behavior of the individual pesticides is best shown by the plots in Figures 3 and 4. Common names
for the pesticides are alphabetized on the plots for convenience
of location. The formulations deposited are shown immediately
below the common names. The vertical axes are either 0-15, 0-150
or 0-1500 μg/g depending on the maximum concentration observed for
any one residue. In some cases, such as benomyl and trifluralin
in Figure 3, no data points are shown in 1977 because these pesticides were either not measured, as is the case for trifluralin, or
the analytical schemes had not yet been devised as is the case for
benomyl. Also no data points are shown for bensulide in 1977
because none had been dumped into the pit.

Extracts of samples from the Horticulture pit were also analyzed for alachlor, carbaryl, glyphosate, dicofol, MCPP, paraquat,
penoxalin and 2,4,5-T. They were either not detected or, if
detected, concentrations were <1 ppm. No plots of these results
are included in this report. In addition, plots are not included
for several pesticides where only results from 1979 were available. These pesticides, with their average ppm concentrations in
parentheses, were: 2,4-D (40), dicamba (3), metribuzin (25),
naptalam (60) and tetrachloroterephthalic acid (30).

All data points in Figures 3 and 4 are connected by straight
lines. The graphical data thus give a quick visual representation
of probable build-up only. They are not representations of degradation per se. The probable degradation may be deduced by careful
inspection of these data and the deposition data given in Table
VI. However, mathematical adjustments of the concentrations based
on deposition factors were unsuccessful for deducing degradation
because of the random nature of the sampling and deposition dates,
incomplete containment and irreversible adsorptions. The analytical uncertainties, due to the highly heterogeneous nature of these
pits, also argue against adjusting the analytical data with deposition factors in order to deduce whether degradation has occurred.

In general most of the pesticide residues are immobilized on
soil particles, although in some cases, such as alachlor, atrazine, benomyl, diphenamid, EPTC, naptalam and propachlor, significant amounts were found in the liquid phase. This presence in the
liquid phase, which is primarily aqueous, can cause environmental
problems if the liquid contents are lost due to leakage.

Surrounding Water Analyses. Water samples were taken from a tile
system immediately surrounding the Horticulture pit and from sites
remote from the location of the disposal pit. These water samples
were analyzed to ascertain pesticide leakage from the pit and
contamination of the surrounding ground water over that which

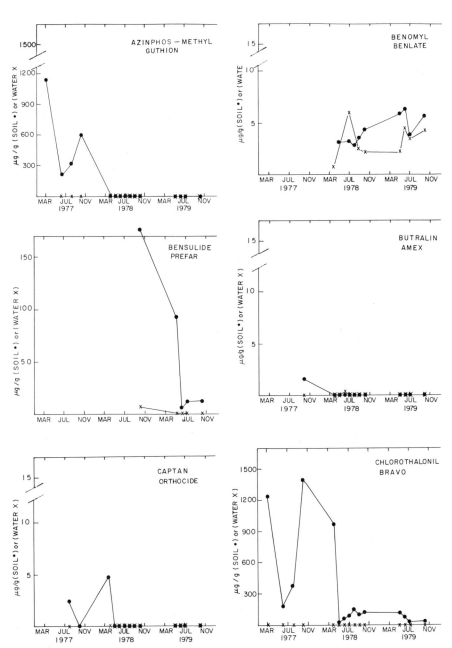

Figure 3.   Plots of the pesticide concentrations in samples of soil (o) and water (X) taken from the Horticulture pit during 1977-1979.

Continued on next page

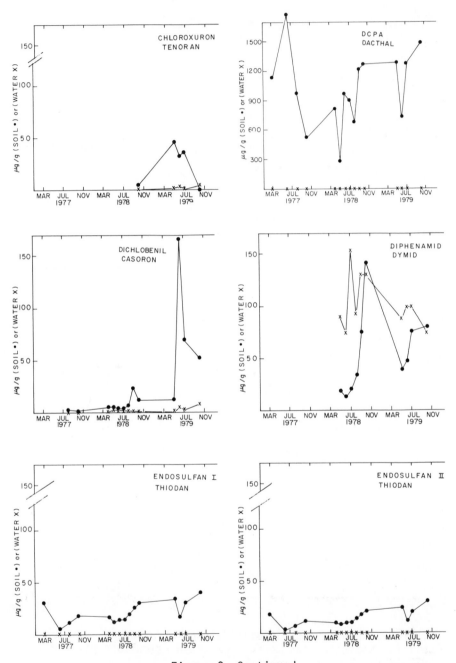

Figure 3. Continued

Continued on next page

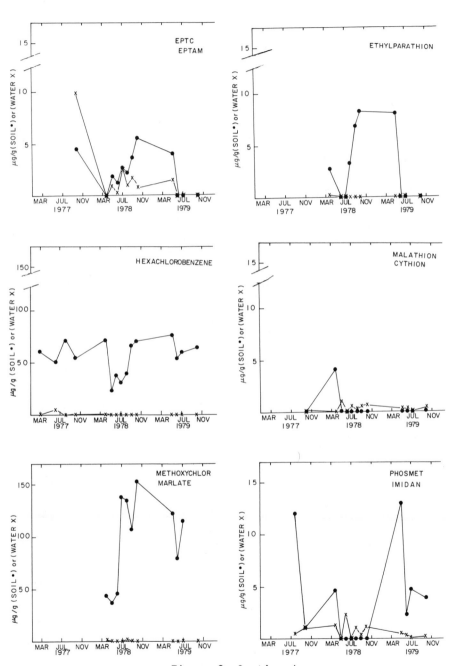

Figure 3. Continued

Continued on next page

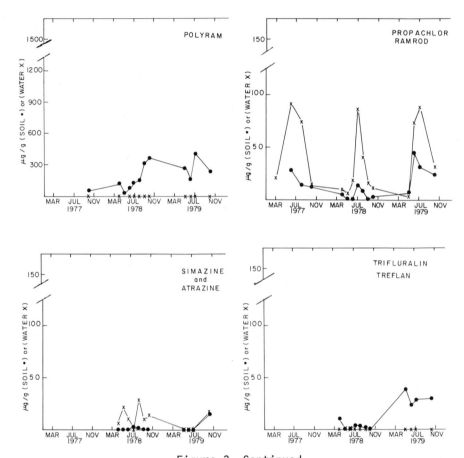

Figure 3. Continued

Table VII. Pesticides in Water Samples Near the Horticulture Pit

Pesticide	mg/L Water					
	Adjacent to Pit			Hort. Pond		Various[†]
	Apr 1978	Jun 1979	Sep 1979	Apr 1978	Jun 1979	Jul 1978
Alachlor	*	*	*	*	0.001	0.0017
Chlorothalonil	*	0.020	*	0.010	*	*
2,4-D	-	0.006	0.010	-	0.0015	-
DCPA[**]	0.002	0.002	0.004	0.004	0.001	0.002
Dicamba	-	0.0006	*	-	0.0002	-
Dichlobenil	*	0.010	*	*	*	*
Endosulfan I	0.001	*	*	0.0001	*	*
Hexachlorobenzene	*	*	*	0.0001	*	*
Propachlor	*	*	0.060	0.005	0.0001	*
Simazine + Atrazine	*	*	0.050	*	0.007	*
Trifluralin	0.010	0.040	*	*	*	*
Tetrachloroterephthalic Acid	-	0.041	0.33	-	0.005	-

*Not detected

⁻Not measured

†Average concentration of samples collected at twelve remote
locations at the Horticulture Farm.

**DCPA measured Aug. 1978 and Apr. 1979 was 0.001 and 0.006 PPM
respectively in the water adjacent to the pit.

occurred from other probable sources. The analytical results are
given in Table VII where only those pesticides from Table VI which
were found above their detection limit are listed. The data under
the columns labelled adjacent to pit represent the amounts of the
pesticides found in the tile immediately surrounding the pit. Any
large amount of a soluble pesticide observed here would be indicative of leakage from the pit and possible contamination of surrounding ground water. Repeat analyses over a relatively long
time period are necessary for proper interpretation of these data
since inadvertent contamination may distort the analyses for any
short time period. Nevertheless, the periodic analyses of the
ground water from the tile surrounding the pit are useful as a
first indicator of a problem in the operation of the disposal
pit.

The Horticulture pond designated in Table VII is located

about 300 m from the disposal site and it receives seepage and
run-off from all surrounding areas. Water from this pond has
measurable amounts of several pesticides but this contamination is
related to run-off from surrounding farm land rather than contamination from the disposal pit. The chemicals and the amounts observed here in the early spring of 1978 and 1979 are typical for
surface water from farm ponds in Central Iowa. The data are not
related to the amounts of the pesticides deposited in the disposal
pit.
     Water was also taken from a well located within 15 m of the
disposal pit and no pesticides were found at the very low detection limits of 1 to 10 ng/L.

Surrounding Air Analyses. The results of the analyses of the air
taken above the Horticulture pit are given in Table VIII for the
spring and fall of 1979. Air samples were also taken upwind from
the disposal pit and most pesticides were not detected or if detected were at their detection limit of ~0.1 ng/L. Volatilization
from the pit may be assumed whenever there is an appreciable increase in pesticide concentration in air over the pit relative to
an upwind sample taken at the same time.

Table VIII. Pesticides in Air Above Horti-
             culture Pit

Pesticide	ng/L Air	
	6/28/79	9/21/79
DCPA	1.1	1.2
Dichlobenil	2.5	1.1
Endosulfan I	0.3	0.2
Endosulfan II	<0.1	<0.1
Hexachlorobenzene	0.3	0.3
Methoxychlor	<0.1	<0.1
Trifluralin	*	0.5

*not detected

     Sufficient samples at the proper locations were not taken to
allow for accurate estimates of the vaporization losses for those
pesticides which could be proven to be emanating from the disposal
pit. However the low concentrations measured above the pit suggest that losses by volatilization are not a source of significant
contamination of surrounding air. The pit therefore is a source
of air contamination due to volatilization but the amount is negligible given the normal dilution effect for point sources of air
pollution.

Surroundings Summary - Horticulture. From interpretation of the
above results for water and air surrounding the Horticulture pit,
very little contamination can be attributed to the operation of
this pit disposal system.

Agronomy Pit

The accumulations of the amounts of pesticides deposited in the
Agronomy pit are given in Table IX. This pit was constructed in

Table IX. Disposal Accumulations at the Agronomy
Pit in Grams of Active Ingredient (AI)

Formulation Deposited	Grams AI at Listed Dates	
	Nov 1, 1978	Oct 1 1979
Aatrex	17488	21393
Amiben	1898	2511
Banvel	45	272
Basagran	1249	2612
Bladex	6973	13452
Cygon	123	212
2,4-D amine	558	1175
Dual	1021	1171
Eradicane	1807	3668
Furloe	2561	5853
Lasso	31031	53634
Cythion	427	427
Paraquat	300	800
Ramrod	11186	11186
Roundup	4087	5086
Sencor	3722	4675
Sutan	4654	4745
Tolban	91	91
Treflan	10138	11409
TOTAL =	99359	144372

4540 G of an unspecified number of pesticides
were also dumped on or about Feb 1, 1978. The
total grams and the listings are thus absolute
minimums. In addition to the listed pesticides
8508 g of Bexton, Prowl, Ravage, Sevin and
Surflan were deposited but not analyzed.

the late fall of 1977 and first used in February of 1978. The pit design precluded getting soil samples; so only liquid samples were taken for analyses. Even though data already discussed for the Horticulture pit show most of the pesticide concentration to be associated with soil samples, these analyses are not necessary for the assessment of the utility of a pit disposal system. Loss of the solid contents of a pit is highly improbable. However, some or all of the liquid contents can be lost if cracks develop in the pit liner. Therefore the amounts of pesticides in the liquid phase should be measured regularly to account for possible losses due to leakage and to estimate the maximum contamination from the most adverse leakage problem. In practice, a rupture should be repaired before any significant escape of pesticides occurs because of leakage of the liquid contents. Because some small amount of all pesticides are present in the liquid phase, the regular sampling and analyses of this phase is also desirable for determining whether a determental build-up is occurring. The interpretation of the liquid monitor data also provides evidence of degradation.

The Agronomy pit is also an uncontrolled disposal system so the analytical data are perturbed by the same variables of amount, timing, degradation and losses as described in the discussion of the Horticulture pit.

Liquid Analyses. The pesticide residues in the liquid samples taken from the Agronomy pit are given in the graphical plots in Figure 4 where the amounts present at each sampling date are plotted separately for each pesticide. The vertical axes are either 0-15, 0-150 or 0-1500 µg/g of liquid sample depending on the maximum concentration observed for any one residue in the liquid phase. The pesticide plots are ordered alphabetically for ease of location. The formulations are shown immediately below the common names on the plots.

Extracts of samples from the Agronomy pit were also analyzed for chlorambem, glyphosate, malathion, metolachlor and perfluralin. They were either not detected or, if detected, concentrations were <1 ppm. Plots are not included for several pesticides where only results from 1979 were available. These pesticides, with their average ppm concentrations in parentheses, were: bentazon (150), 2,4-D (30), dicamba (10), paraquat (20) and 2,4,5-T (40).

As with the Horticulture pit data, the connection of these data points with straight lines is not a valid representation of degradation because of the uncontrolled manner in which the pesticides are deposited. The plots do however give a quick visual representation of possible build-up of any pesticide. Probable degradations may be estimated by comparing the deposition information in Table IX with the analytical data in Figure 4.

In some instances, as with 2,4,5-T, measurable amounts were present; yet no records of deposition were available. These

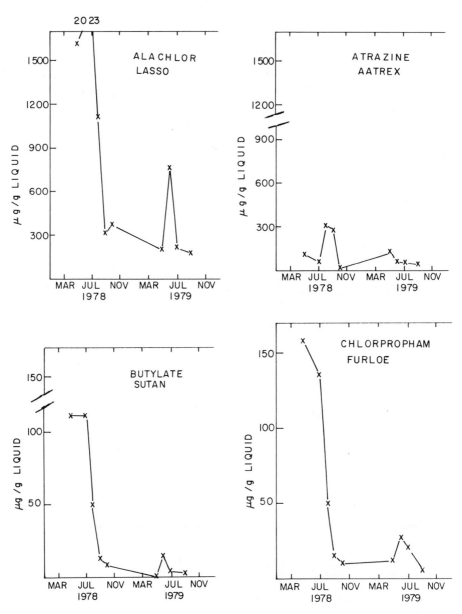

Figure 4.   Plots of the pesticide concentrations in liquid samples taken from the Agronomy pit during 1978-1979.

Continued on next page

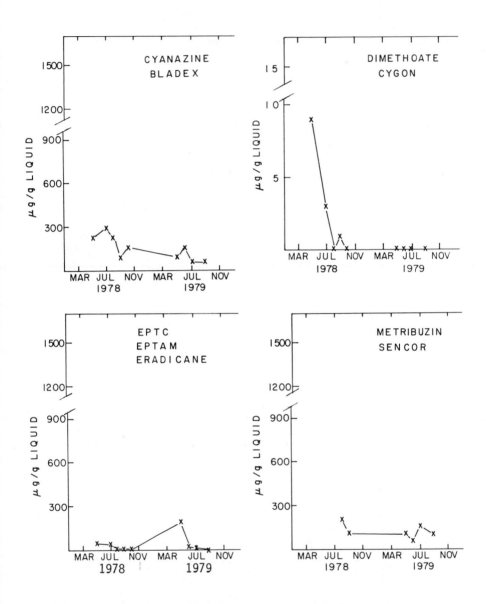

Figure 4. Continued

Continued on next page

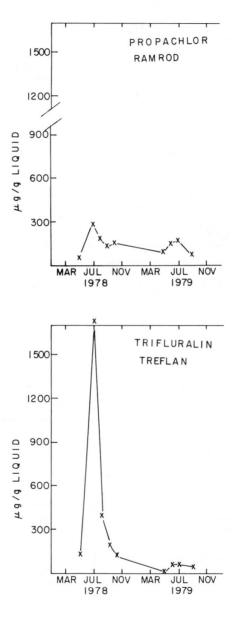

Figure 4. Continued

observations focus sharply on the practical problem of providing completely accurate disposal data.

Surrounding Water and Soil Analyses. The results of the analyses of ground water samples taken near the Agronomy pit are given in Table X. No drainage tile surrounds this plastic lined pit, so the May 31, June 15 and June 28 samples were taken from the ground water seepage into 10 cm holes augered at each of the four outside corners of the pit. These data suggest some seepage of chemicals

Table X. Herbicides in Ground Water Immediately Adjacent to the Agronomy Pit

	mg/L Water		
Herbicide**	May 31 1979	Jun 15 1979	Jun 28 1979
EPTC	0.03	*	0.01
Butylate	0.11	*	0.04
Chlorpropham	*	*	<0.01
Atrazine	*	*	0.28
Trifluralin	*	*	*
Alachlor	*	*	0.03
Cyanazine	*	*	<0.01
Metribuzin	0.22	0.31	0.34
Bentazon[†]	2.5	2.8	15
Dicamba[†]	*	*	2.1
2,4-D[†]	0.10	0.06	0.06
2,4,5-T[†]	0.25	0.03	8.8

**Propachlor, Dimethoate and Diphenamid not detected at 0.01 ppm.
*not detected at <0.01 PPM
[†]measured as methyl ester

into the ground water immediately adjacent to the pit but the long range effect of this possible leakage is uncertain. Additional data would be necessary to confirm positively the extent of the chemical losses by the suspected route.

Soil samples taken at the corners of the Agronomy pit were composited and portions were analyzed for pesticide residues. No pesticides were found at the detection limit of 10 μg/g soil.

Surrounding Air Analyses. The results of the analyses of the air taken above the Agronomy pit are given in Table XI for June 28, August 14 and August 24, 1979. The interpretation of these results relative to containment are the same as that discussed for the Horticulture pit.

Table XI. Pesticides in Air Above Agronomy Pit

Pesticides [†]	ng/L Air		
	6/28/79	8/14/79	8/24/79
EPTC	0.6	1.0	4.5
Butylate	0.3	<0.1	1.2
Propachlor	<0.1	<0.1	*
Atrazine	<0.1	*	*
Trifluralin	0.3	0.6	0.8
Alachlor	1.5	0.5	0.9
Cyanazine	<0.1	*	*
Metribuzin	0.1	0.1	0.2

[†]Chlorpropham, Dimethoate and Diphenamid not detected at 0.1 mg/L.
*not detected

   The evidence for pesticide losses due to evaporation is as expected for the more volatile pesticides such as EPTC, butylate, trifluralin and alachlor. Losses of other pesticides due to evaporation from the pit surface are uncertain because similar results were obtained for analyses of samples taken upwind and above the pit.

Surroundings Summary – Agronomy. From interpretation of the above results for soil, water and air surrounding the Agronomy pit, little contamination can be attributed to its operation.

Evidence of Degradation

Many pesticides degrade to polar products that form organic anions in a water matrix. These are sometimes missed due to the diffi-culty in extracting trace amounts of the ionic material from a water matrix. A recently developed anion exchange procedure for isolating acidic compounds from dilute aqueous solutions (10-12) was used for recovering the anionic material from liquid samples collected from the two pits.
   Twenty-five acidic pesticides, most of which resulted from the rapid hydrolysis of the ester formulations, were isolated from the two pits. Thirty-nine substituted aromatic acids and phenols were also isolated from the pits. The source of these components was most likely the degradation of aromatic pesticides and the

aromatic solvents used to produce the liquid formulations. Also, fourteen organophosphorus acids from the hydrolysis of the organophosphorus insecticides were identified along with 19 aliphatic and other miscellaneous acids. These acids may have originated with the emulsifying agents.

The finding of 97 acidic components in the two pits is a good indication that the deposited pesticides are being degraded in addition to being contained.

CONCLUSIONS

1. Sampling is extremely difficult in a large disposal system containing soil, rock and liquids. At very best, a rough estimate only is obtained of the amounts of pesticides that are present. The probable errors are compounded by segregation (concentration differences) in the soil which can be as high as 1000 times. Even segregation in the liquid phase is appreciable being as high as 10 times.

2. Containment appears to be satisfactory in both the concrete and the plastic lined pit.

3. Soil is essential in a disposal system as a source of organisms and food and for containment by adsorption.

4. Rock provides an element of safety but creates sampling problems. Other means, such as a grate, could be employed to prevent accidental contact with the liquid phase by humans and animals.

5. Disposal pits should be designed for convenient sampling of the liquid contents. The soil doesn't move, so contamination will occur exclusively through the liquid and vapor phase. Containment can be estimated by infrequent sampling of the liquid contents and regular sampling and analyses of the surrounding water taken from shallow wells outside the pit and the air taken from above the pit.

ACKNOWLEDGMENTS

Part of this work was performed in the laboratories of the U.S. Department of Energy and supported under contract No. W-7405-Eng-82, Division of the Office of Health and Environmental Research, Office of Energy Research. The assistance of E. Catus, C. V. Hall, R. S. Hansen, T. E. Hazen and H. J. Svec in the administration of EPA contract No. R804533010, (C. Rogers, Project Officer) which provided the major support of this research is appreciated. The cooperation of the staff at Iowa State University involved in pesticide disposal research is acknowledged. These include J. L. Baker, J. M. Chow, P. A. Dahm, R. D. Fish, C. V. Hall, P. A. Hartman, R. P. Nicholson, J. T. Pesek, G. A. Schuler, D. W. Staniforth and F. D. Williams.

LITERATURE CITED

1. Hall, C.; Baker, J.; Dahm, P.; Freiburger, L.; Gorder, G.; Johnson, L.; Junk, G.; Williams, F. Safe Disposal Methods for Agricultural Pesticide Wastes, U.S. EPA Report 600/2-81-074, NTIS Accession PB 81 197584 (1981).

2. Junk, G. A.; Richard, J. J.; Grieser, M. D.; Witiak, D.; Witiak, J. L.; Arguello, M. D.; Vick, R.; Svec, H. J.; Fritz, J. S.; Calder, G. V. J. Chromatogr. 1974, 99, 745-762.

3. Woolson, E. A. J. Assoc. Off. Anal. Chem. 1974, 57, 604-609.

4. Richard, J. J.; Junk, G. A.; Avery, M. J.; Nehring, N. L.; Fritz, J. S.; Svec, H. J. Pest. Monit. J. 1975, 9, 117-123.

5. Junk, G. A.; Richard, J. J.; Fritz, J. S.; Svec, H. J. in "Identification and Analysis of Organic Pollutants in Water"; Keith, L. A., Ed.; Ann Arbor Science: Ann Arbor, 1976; Chap. 9.

6. Junk, G. A.; Richard, J. J. in "Identification and Analysis of Organic Pollutants in Air"; Keith, L. H., Ed.; Ann Arbor Science: Ann Arbor, Mich., 1983; (in press).

7. Austin, D. J.; Briggs, G. G. Pestic. Sci. 1976, 7, 201-210.

8. McLeod, H. A.; McCully, K. A. J. Assoc. Off. Anal. Chem. 1969, 52, 1226-1230.

9. King, R. R. J. Agric. Food Chem. 1978, 26, 1460-1463.

10. Richard, J. J.; Fritz, J. S. J. Chromatog. Sci. 1980, 18, 35-38.

11. Junk, G. A.; Richard, J. J. in "Advances in the Identification and Analysis of Organic Pollutants in Water"; Keith, L. H., Ed.; Ann Arbor Science: Ann Arbor, Mich., 1981; Chap. 19.

12. Junk, G. A.; Richard, J. J. Water Qual. Bull. 1981, 6, 40-43.

RECEIVED February 13, 1984

Disposal of Pesticide Wastes in Lined Evaporation Beds

W. L. WINTERLIN, S. R. SCHOEN, and C. R. MOURER

Department of Environmental Toxicology, University of California at Davis, Davis, CA 95616

Ten lined evaporation beds for disposing of pesticide wastes from used pesticide containers and application equipment were monitored over a two-year period for possible buildup and decay of deposited pesticides. The evaporation beds had been in operation for many years prior to sampling and were distributed throughout the state with geographical and climatical differences. All the beds had the same basic design in that the pesticides washings were supplied to the bed underneath the soil surface through leach lines. There were some differences in design of the various beds with the most significant being some beds had hydrated lime incorporated into the soil while others did not. Two of the beds had large storage tanks prior to transport of the washings into the bed. There was also a large difference in the quantity and type of pesticides deposited in some of the beds. The soils in the beds were sampled in quadrants at 0-1, 1-6, and 6-12 inch depths. Air samples along the edge of the bed were also taken during the second year of sampling. Conclusions from these investigations showed that (1) the beds do not generally build up high levels of pesticides and are effective in containing as well as degrading the pesticide without excessive exposure via air vapor; (2) pesticides generally rise to the surface of the bed where they can be degraded by photochemical, chemical, and biological forces as well as be distributed via air vapor and (3) the amendment of the beds with lime may be an important factor in the degradation of some pesticides.

0097-6156/84/0259-0097$06.00/0

Wastes created by rinsing of used pesticide containers and application equipment create a unique disposal problem. There are large volumes of waste being generated containing dilute toxic material that must be properly disposed of, and the waste generators are not centrally located. Using aerial applicators as an example, it has been estimated that 10 to 60 gallons of wastewater containing 100 to 1000 ppm of pesticides are generated per day per plane (1). Storing the wastewater and having it hauled to a disposal site can be expensive, impractical, and in the future may be prohibitive for some pesticides. This creates a need for an economical, small scale, on-site disposal system.

For any on-site disposal system, there are at least four criteria that should be met to decrease the chance of environmental contamination and either eliminate or decrease the need for disposal at a permanent facility. First, the disposal system would need to concentrate the large volume of waste to increase ease of handling and permanent disposal. Secondly, the disposal system would be required to contain the pesticide wastes to avoid environmental contamination or risks to human health. Thirdly, the system should prevent toxic levels from vaporizing into the air. Finally, the disposal system should degrade and detoxify the pesticide wastes in the containment, particularly if buildup of the wastes to toxic levels is a potential problem.

Evaporation Bed Experimental Design

The University of California field stations have dealt with dilute pesticide waste disposal on an experimental basis by using lined soil evaporation beds. The beds typically are 20 x 40 x 3 ft pits lined with a butyl rubber membrane and back filled with 12 to 18 inches of sandy loam soil. Figure 1 is a cross secton of such a bed. Used containers and spray equipment are washed on an adjacent concrete slab; the wastewater drains into a sedimentation box for trapping particulates, followed by a distribution box in the bed. From the distribution box, the dilute pesticide solutions run underneath the soil surface through leach lines made of 4 inch perforated PVC pipe. The system is designed so that water moves up through the soil by capillary action and evaporates off the surface.

In order to reduce the possibility of the beds inadvertently filling up with water, three design features were added. Most of the beds were covered with a corregated fiberglass roof. In addition to keeping off rainfall, the fiberglass still allows sunlight to penetrate through to heat the bed's surface and increase evaporation. For additional flood control, one half of the field stations have roofs over the wash slabs. Two of the stations use tanks to temporarily store excessive amounts of rinsings during high use periods.

In order to increase the degradation rate of certain pesticides, in most of the beds approximately 1 ton of hydrated lime [Ca(OH)$_2$] was incorporated into the soil. However, not all beds had the same amount of lime nor was the lime incorporated into the bed in the same manner. One evaporation bed, for example, had applied the lime as a thin film on top of the bed without any apparent effort to mix it into the soil beyond raking. The use of lime was to increase the soil pH to accelerate the hydrolysis of organophosphate and carbamate pesticides.

The bed size needed for any given user may be calculated once the region's pan evaporation rate (inches/month) and wastewater volume addition rate (gal/month) are known. The equation is:

$$\text{area required (ft}^2\text{)} = \frac{\text{usage (gal/mo)}}{0.8[\text{pan evap rate (in/mo)}](0.625)}$$

where 0.8 is the reduced evaporation rate from sandy loam soil compared to an open pond and 0.625 is a conversion between gallons and inches/sq ft. A safety factor of 100% (using twice the bed size required) has been suggested (2). Because of the importance of evaporation rate in reducing the bed size requirement, evaporation beds could be expected to be more effective in warm, dry climates such as California's.

Once in an evaporation bed, a pesticide can adsorb to a soil colloid, undergo chemical or microbial degradation, or escape from the bed by volatilization. An evaporation bed has the potential advantage over an open pond of decreasing pesticide volatilization while allowing for increased degradation through microbial and soil-catalyzed reactions.

Ten University of California field stations located throughout the state (Figure 2) allow for the study of the evaporation beds under several climatic and design variations (Table I). All the evaporation beds were sampled except the one located farthest away, at the Imperial Valley station. The evaporation beds vary in temperature range and annual precipitation, and in such design considerations as size, soil depth, soil pH, the presence or absence of storage tanks, a roof over the bed or slab area, and whether or not the soil itself has been amended with hydrated lime. The beds also differ in the specific pesticide wastes disposed, and the longevity and intensity of use.

In 1981, a study was undertaken to determine whether the evaporation beds were functioning as originally intended after 4 to 8 years of use. The major concern at that time was to determine if the pesticides were building up in the beds, and, if so, what could be done to correct the problem without the expense of physical removal of the contaminated soil to a Class I or II dumpsite.

Methods

Sampling. Both soil core and air samples were taken periodically from the beds. Four replicate soil samples from each of four quadrants were taken at 0-1, 1-6, and 6-12 in. depths. The 0-1

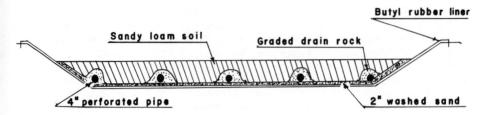

Figure 1. Cross-section of a University of California evaporation bed.

Figure 2. Locations of University of California field stations where evaporation beds are operating.

Table I.   Construction Details of Evaporation Beds Where Samples Were Taken

	South Coast	West Side	Kearney	Lindcove	Deciduous	Davis	Sierra	Hopland	Tulelake
Year constructed	1976	1973	1974	1973	1973	1975	1977	1977	1973
Roof over bed	Yes	Yes	Yes	Yes	Yes	Yes	Yes	Yes	No
Roof over slab	Yes	No	Yes	No	Yes	No	Yes	Yes	No
Storage tanks	No	No	Yes	No	No	Yes	No	No	No
Soil depth	12-15"	12-15"	4-10"	12-18"	6-12"	18-24"	18-24"	12-15"	12-15"
Lime	Yes	No	No	Yes	Yes	No	Yes	Yes	Yes
Use	Heavy	Heavy	Very Heavy	Heavy	Moderate	Moderate	Low	Low	Moderate

in. samples were taken with a small spatula from a 10 cm x 10 cm
area, while the 1-6 and 6-12 in. samples were taken using a 2 in.
diameter soil core sampler. Air samples were taken at the downwind edge of each bed by running a high volume air sampler with a
quartz fiber filter backed with 120 mL of XAD-4 resin for 2 hr.

Extraction and Analysis. Soil samples (10 g dry weight) were
extracted with 200 mL ethyl acetate for 30 sec using a Tissuemizer
ultrasonicator. To extract phenoxyacids, the samples were acidified with HCl and re-extracted. Quartz fiber filters and XAD-4
resin from air samples were soxhlet extracted overnight using 350
to 500 mL ethyl acetate. A 5 g equivalent aliquot of the soil
sample extract, or half of the air sample extract, was evaporated,
derivitized with diazomethane, and brought to the proper concentration for analysis.

Samples were analyzed by gas chromatography using nitrogen-
phosphorus, flame photometric (sulfur mode), and Dohrmann
microcoulometric (halogen mode) detectors. The chromatographic
conditions are listed in Table II. Using the selective detectors,
pesticides were tentatively identified by comparison with standard
retention times, and quantified using standard curves. They were
then confirmed by using a capillary GC/MS/data system. Over 100
common pesticide standards were used for screening. It should be
noted that only gas chromatographable pesticides were analyzed.
Minimum detectable levels varied depending on the number and
amount of pesticides found in the samples as well as the relative
response of the pesticide to the detectors. Generally detectable
levels ranged from 0.1 ppm to 1.0 ppm. Metabolites were not
determined.

Sampling of the air on the side downwind from the evaporation
beds was conducted during 1982 using XAD-4 macroreticular resin.
A high volume Staplex air sampler with a vacuum capacity of 24
L/min was placed horizontally on top of the berm facing towards
the bed and operated for 2 hr. The air sampling procedure including the preparation of the resin and extracting the captured
pesticide vapors has been reported by Wehner et al. (3).

pH Determination. pH measurements were made using the procedure
of Smith and Atkinson (4). Distilled water (50 mL) was added to
20 g of the dry soil and thoroughly mixed. pH determinations were
then made on a Corning Model 135 pH/ion meter and recorded.

Results and Discussion

The evaporation bed samples were generally a complex mixture of
many compounds and good records of inputs to the beds were not
always kept. Also some of the beds were nonuniform in their
distribution of the pesticides as well as soil depth. All beds
had been packed with sandy loam and when sandy loam was fortified
with 11 of the most frequently found pesticides, the average

Table II. Chromatographic Conditions Used for Evaporation Bed Samples

Gas Chrom	Detector	Column Packing	Temp. (°C)			Gas flows (cc/min)				
			Oven	Inj	Det	N_2	H_2	He	O_2	Air
Hewlett Packard 5710A	N/P	(1) 6% OV101 on Chromsorb W HP 80/100	200	210	250	--	4	30	--	75
		(2) 180 cm 4.5% OV225 50 cm 4.0% OV101 on Supelcoport 80/100	200	210	250	--	4	25	--	75
Tracor MT550	FPD	12% SE-30 on Gas Chrom Q, 80/100	210	210	190	50	45	--	5	95
Tracor MT220	Dohrmann Coulometer Cl cell	6% DC-200 and 1.5% QF-1 on Gas Chrom Q 80/100	150-175	200	--*	30	--	--	80	--
Finnigan 3200 E**	EI	DB-5 25 m × 0.25 mm fused silica	150-200 8°/min	210	--	--	--	33	--	--

* Detector at ambient temperature with the pyrolysis oven at 660°C and coulometer set at 200 ohms.

** 6000 data system. Split ratio 30:1; e^- multiplier 1600 ev; e^- source 70 ev recoveries ranged between 89 and 100% (Table III). Table IV shows averaged total pesticide residue levels and the standard deviation found in the top inch of soil taken from the four quadrants in 6 of the more heavily used beds during the two sampling periods in 1982. All of the beds were nonuniform in their distribution of the pesticides with the bed at Lindcove containing the highest pesticide levels and also the greatest nonuniformity. However, the number and quantity of pesticides found within each quadrant were relatively the same proportions at the three depths with those in the other quadrant. Therefore, the degree of nonuniformity was apparently due to distribution of the pesticides into the

Table III. Method Recoveries of Pesticides Most Commonly Found in Evaporation Beds

		% Recovery				
Pesticide	Fortification level (ppm)	Rep I	Rep II	Rep III	Average Recovery	S.D.
Atrazine	5.0	91.9	91.9	94.4	92.7	1.44
Chlorpyrifos	5.0	92.8	91.2	94.9	92.9	1.85
DACTHAL	5.0	93.5	91.5	89.0	91.3	2.25
DEVRINOL	5.0	100.0	100.0	100.0	100.0	0.00
Diazinon	5.0	89.6	87.2	91.2	89.3	2.01
Endosulfan I	5.0	95.5	95.5	97.5	96.2	1.15
Endosulfan II	5.0	92.5	90.5	95.5	92.8	2.52
Parathion	5.0	95.6	91.2	96.3	94.4	2.76
Simazine	5.0	96.2	94.9	97.5	96.2	1.30
Trifluralin	5.0	92.4	90.5	92.4	91.8	1.10
2,4-D	10.0	90.0	94.5	85.9	90.1	4.30

Table IV. Average Soil Surface Residues (0-1") Exceeding 1 ppm in Evaporation Beds Located at Six Field Stations Throughout California in 1982

	Spring		Summer	
	$\bar{x}$ (ppm)	S.D.	$\bar{x}$ (ppm)	S.D.
South Coast	21.06	37.80	--	--
West Side	289.2	439.7	398.5	663.3
Kearney	100.8	122.2	141.4	225.8
Lindcove	2,031	5,490	7,465	14,666
Davis	88.85	160.8	50.60	74.01
Tulelake	40.66	87.51	270.4	454.8

beds. During the winter and early spring months, many of the beds contained standing water due to the relatively low temperatures, foggy climate, lack of wind, heavy rains and heavy use from dormant and other spray applications. Beds were never known to overflow, however. Since air movement is such an important factor in evaporation, the installation and operation of a powered fan may prove beneficial in future installations, particularly during the winter months.

Even though the number of pesticides surveyed amounted to more than 100, only 46 were actually detected (Table V). From records that were kept as well as unidentified GLC peaks, other pesticides were deposited into the beds but either were not detected by the gas chromatography procedure, were present in undetectable levels or were unknown or unregistered compounds. Tables VI to XI show the average accumulated pesticide concentrations from each of the six field stations having the heaviest use of the evaporation beds.

In spite of these many variables, there did appear to be some major conclusions from the study. First, with the exception of the Lindcove bed, the evaporation beds had low soil pesticide levels for a disposal facility of this magnitude. Table XII is a summary of those pesticides found in each bed which had average residues in excess of 5 ppm during the two-year sampling period.

As a general rule, the more heavily used beds tended to have higher pesticide residues. The evaporation beds at Deciduous, Sierra Foothill, and Hopland field stations, which were not used

Table V. Pesticides Detected in University of California Evaporation Beds Using Gas Chromatography with Specific Element Detectors

Pesticide	Detector(s)	Pesticide	Detector(s)
Atrazine	Coul, NP	DEVRINOL	NP
Azinphosmethyl	FPD, NP	Diuron	Coul
Benzulide	FPD, NP	2,4-D, methyl ester	Coul
Benthiocarb	Coul, NP	Dicofol	Coul
BLADEX	Coul, NP	Dimethoate	Coul, FPD
Carbaryl	NP	Disulfotan	FPD, NP
Chlorpyrifos	Coul, FPD, NP	Endosulfan	Coul, FPD
CIPC	Coul, NP	EPTC	FPD, NP
DACTHAL	Coul	Ethion	FPD, NP
DEF	FPD, NP	Malathion	FPD, NP

Continued on next page

Table V. Continued

Pesticide	Detector(s)	Pesticide	Detector(s)
Methidathion	FPD, NP	Pirimicarb	NP
Methyl parathion	FPD, NP	Prometryn	FPD, NP
MCPA, methyl ester	Coul	Propachlor	Coul, NP
MOBAM	FPD, NP	Propazine	Coul, NP
Molinate	FPD, NP	Propoxur	NP
Nitrofen	Coul, NP	RONEET	FPD, NP
OMITE	FPD	Simazine	Coul, NP
Parathion	FPD, NP	Terbacil	Coul, NP
PCNB	Coul, NP	Tedion	Coul, FPD
Pebulate	FPD, NP	Trifluralin	NP
Phorate	FPD, NP	Vernolate	FPD, NP
Phosvel	Coul, FPD, NP	ZYTRON	Coul, FPD, NP
Picloram	Coul, NP		

Table VI. Pesticide Soil Concentration in Relation to Depth from
           Averaging Four Sampling Areas in Evaporation Bed
           Located at South Coast Field Station

	0-1"		PPM 1-6"		6-12"	
Pesticide	3/20/81	4/30/82	3/20/81	4/30/82	3/20/81	4/30/82
Azinphosmethyl	1.0	n.d.	n.d.	n.d.	n.d.	n.d.
Bensulide	n.d.	7.2	n.d.	n.d.	n.d.	n.d.
Chlorpyrifos	14	1.0	n.d.	n.d.	n.d.	n.d.
2,4-D	27	32	0.15	n.d.	0.30	n.d.
DEVRINOL	5.3	17	n.d.	3.9	1.9	3.2
MCPA	3.3	n.d.	n.d.	n.d.	n.d.	n.d.
MOBAM	n.d.	9.7	n.d.	8.6	n.d.	n.d.
Molinate	n.d.	1.4	n.d.	n.d.	n.d.	n.d.
Prefar	19	n.d.	n.d.	n.d.	n.d.	n.d.
Propoxur	n.d.	6.1	n.d.	n.d.	n.d.	n.d.
Simazine	2.4	7.6	n.d.	4.0	n.d.	2.0

Table VII. Pesticide Soil Concentrations in Relation to Depth from Averaging Four Sampling Areas in Evaporation Bed Located at West Side Field Station

PPM

	0-1"				1-6"				6-12"			
	1981		1982		1981		1982		1981		1982	
Pesticide	3/17	7/7	4/29	9/17	3/17	7/7	4/29	9/17	3/17	7/7	4/29	9/17
Bensulide	n.d.	n.d.	499	690	n.d.	n.d.	332	182	n.d.	n.d.	24	38
Carbaryl	30	130	n.d.	n.d.	n.d.	n.d.	n.d.	n.d.	n.d.	n.d.	n.d.	n.d.
Chlorpyrifos	198	396	217	127	243	236	173	60	72	4.4	16	15
CIPC	270	425	478	n.d.	253	194	196	n.d.	73	6.8	17	n.d.
DACTHAL	488	1867	956	1383	505	205	516	202	148	4.6	956	14
DEF	102	529	221	88	263	279	204	33	86	7.9	16	9.0
DEVRINOL	473	581	403	713	103	72	46	32	31	5.5	14	12
Disulfoton	n.d.	n.d.	3.9	n.d.	n.d.	n.d.	1.8	n.d.	n.d.	n.d.	n.d.	n.d.
Endosulfan	n.d.	n.d.	69	n.d.	n.d.	n.d.	n.d.	n.d.	n.d.	n.d.	n.d.	n.d.
Ethion	n.d.	n.d.	n.d.	n.d.	n.d.	n.d.	4.6	4.1	n.d.	n.d.	n.d.	n.d.
Malathion	n.d.	n.d.	18	n.d.	n.d.	n.d.	5.3	n.d.	n.d.	n.d.	n.d.	n.d.
Methidathion	n.d.	n.d.	6.1	n.d.	n.d.	n.d.	6.1	n.d.	n.d.	n.d.	n.d.	n.d.
Methyl parathion	47	47	n.d.	n.d.	n.d.	n.d.	217	n.d.	n.d.	n.d.	62	n.d.
MOBAM	n.d.	n.d.	435	n.d.	n.d.	n.d.	n.d.	n.d.	n.d.	n.d.	n.d.	n.d.
OMITE	n.d.	n.d.	780	1039	n.d.	n.d.	714	344	n.d.	n.d.	1.8	47
Phorate	80	45	39	29	13	12	13	13	3.5	n.d.	n.d.	8.9
RONEET	45	245	86	115	133	99	86	46	42	2.7	n.d.	8.9
Simazine	72	255	238	144	141	254	238	27	39	4.8	n.d.	5.6
Trifluralin	412	1835	1029	1063	797	788	332	199	265	16	40	42

Table VIII. Pesticide Soil Concentrations in Relation to Depth from Averaging Four Sampling Areas in Evaporation Bed Located at Kearney Field Station

PPM

Pesticide	0–1"				1–6"				6–12"			
	1981		1982		1981		1982		1981		1982	
	Sprg*	7/7	4/28	9/16	Sprg*	7/7	4/20	9/16	Sprg*	7/7	4/20	9/16
Atrazine	—	3.0	n.d.	22	—	n.d.	n.d.	n.d.	—	n.d.	n.d.	n.d.
Chlorpyrifos	—	n.d.	8.4	6.7	—	n.d.	n.d.	n.d.	—	n.d.	n.d.	n.d.
DACTHAL	—	67	44	65	—	n.d.	n.d.	n.d.	—	1.4	n.d.	n.d.
DEVRINOL	—	260	271	582	—	3.6	3.2	n.d.	—	7.4	1.3	n.d.
Diazinon	—	n.d.	2.6	37	—	n.d.	n.d.	n.d.	—	n.d.	n.d.	n.d.
Endosulfan	—	96	155	113	—	n.d.	n.d.	n.d.	—	1.5	n.d.	n.d.
Eptam	—	n.d.	n.d.	62	—	n.d.	n.d.	n.d.	—	14	n.d.	n.d.
Nitrofen	—	306	n.d.	n.d.	—	n.d.	n.d.	n.d.	—	n.d.	n.d.	n.d.
Picloram	—	n.d.	92	n.d.	—	1.0	n.d.	8.5	—	n.d.	17	3.1
Simazine	—	370	307	495	—	n.d.	n.d.	3.0	—	n.d.	n.d.	0.9
Trifluralin	—	4.5	6.1	229	—	n.d.	n.d.	n.d.	—	n.d.	n.d.	n.d.
Molinate	—	n.d.	n.d.	7.3	—	n.d.	n.d.	n.d.	—	n.d.	n.d.	n.d.
Pebulate	—	n.d.	n.d.	6.9	—	n.d.	n.d.	n.d.	—	n.d.	n.d.	n.d.

* Bed was flooded, preventing sample collection.

Table IX.   Pesticide Soil Concentrations in Relation to Depth from Averaging Four Sampling Areas in Evaporation Bed Located at Lindcove Field Station

	0-1"				PPM 1-6"				6-12"			
	1981		1982		1981		1982		1981		1982	
Pesticide	3/18 **	7/6 *	4/25	9/16	3/18 **	7/6 *	4/28	9/18	3/18 **	7/6 *	4/28	9/16
Atrazine	5.0	n.d.	1,080	1,237	--	--	305	20	--	--	14	3.0
Carbaryl	167,600	967	n.d.	n.d.	--	--	n.d.	n.d.	--	--	n.d.	n.d.
Chlorpyrifos	339	n.d.	102	124	--	--	261	9.5	--	--	109	2.2
DEVRINOL	n.d.	n.d.	n.d.	16,068	--	--	n.d.	8.7	--	--	n.d.	6.9
Dimethoate	756	3.3	362	n.d.	--	--	n.d.	n.d.	--	--	n.d.	n.d.
Ethion	250	50	n.d.	324	--	--	43	36	--	--	7.2	3.1
Methidathion	108	19	n.d.	n.d.	--	--	n.d.	n.d.	--	--	n.d.	n.d.
Methyl parathion	n.d.	n.d.	n.d.	366	--	--	n.d.	n.d.	--	--	n.d.	n.d.
OMITE	n.d.	n.d.	171	12,869	--	--	n.d.	414	--	--	n.d.	111
Parathion	56	22	n.d.	5,945	--	--	2.1	8.7	--	--	n.d.	0.9
Pebulate	n.d.	n.d.	n.d.	21,741	--	--	n.d.	n.d.	--	--	n.d.	n.d.
Prosvel	34	n.d.	n.d.	n.d.	--	--	n.d.	n.d.	--	--	n.d.	n.d.
Propachlor	n.d.	n.d.	44	n.d.	--	--	58	n.d.	--	--	2.6	n.d.
Simazine	16,380	417	9,678	3,807	--	--	305	48	--	--	125	4.5
Terbacil	1,128	n.d.	n.d.	n.d.	--	--	n.d.	n.d.	--	--	n.d.	n.d.
Trifluralin	n.d.	n.d.	2,784	1,261	--	--	62	11	--	--	75	7.8

* Only one sampling area in 1981

** Due to saturation with water, only a 0-6" sample was taken but not at the same location as the other 1981 sample

Table X. Pesticide Soil Concentrations in Relation to Depth from Averaging Four Sampling Areas in Evaporation Bed Located at U.C. Davis

Pesticide	0-1"				1-6" PPM				6-12"			
	1981		1982		1981		1982		1981		1982	
	Sprg*	7/21	5/12	10/14	Sprg*	7/21	5/12	10/17	Sprg*	7/21	5/12	10/14
Atrazine	—	210	146	95	—	12	4.7	4.9	—	11	2.8	9.4
BLADEX	—	335	320	130	—	20	8.0	5.9	—	16	3.8	2.1
Dimethoate	—	n.d.	2.8	n.d.	—	n.d.	n.d.	n.d.	—	n.d.	n.d.	n.d.
Propazine	—	n.d.	2.5	1.8	—	n.d.	n.d.	n.d.	—	n.d.	n.d.	n.d.
Simazine	—	62	56	21	—	3.3	2.7	1.4	—	2.8	1.9	n.d.
Trifluralin	—	n.d.	n.d.	n.d.	—	n.d.	8.0	5.2	—	n.d.	5.0	2.8

* No samples taken

Table XI. Pesticide Soil Concentrations in Relation to Depth from Averaging Four Sampling Areas in Evaporation Bed Located at Tulelake Field Station

PPM

Pesticide	0-1"				1-6"				6-12"			
	1981		1982		1981		1982		1981		1982	
	6/23	10/1	5/7	9/8	6/23	10/1	5/7	9/8	6/23	10/1	5/7	9/8
Chlorpyrifos	1	n.d.	n.d.	18	n.d.	n.d.	n.d.	n.d.	n.d.	n.d.	n.d.	n.d.
2,4-D	11	4.5	n.d.	n.d.	1.6	2.7	n.d.	n.d.	5.6	2.5	n.d.	n.d.
DEVRINOL	n.d.	n.d.	n.d.	n.d.	n.d.	n.d.	n.d.	n.d.	n.d.	n.d.	22.7	n.d.
Diazinon	n.d.	n.d.	n.d.	8.6	n.d.	n.d.	n.d.	n.d.	n.d.	n.d.	n.d.	n.d.
Disulfoton	n.d.	n.d.	n.d.	n.d.	n.d.	n.d.	n.d.	n.d.	2.1	0.50	1.9	n.d.
Endosulfan	n.d.	n.d.	n.d.	*	n.d.	n.d.	n.d.	1.3	103	163	54	273
Molinate	n.d.	n.d.	n.d.	248	n.d.	n.d.	n.d.	28	n.d.	n.d.	n.d.	9.7
Nitrofen	39	0.9	34	1268	n.d.	n.d.	8.1	11	n.d.	n.d.	26	44
Parathion	14	6.4	3.0	207	0.88	n.d.	n.d.	n.d.	25	14	4.6	17
PCNB	375	158	123	90	388	138	132	98	1252	815	2230	828
Phorate	n.d.	n.d.	n.d.	*	n.d.	n.d.	n.d.	n.d.	22	12	17	3.7
Pirimicarb	n.d.	n.d.	1.5	n.d.	n.d.	n.d.	1.8	n.d.	n.d.	n.d.	21	n.d.
Prometryn	1.7	1.9	n.d.	*	4.3	1.0	n.d.	2.0	22	10	7.9	13
Simazine	2.0	1.8	n.d.	n.d.	14	2.7	n.d.	n.d.	41	21	n.d.	n.d.
Trifluralin	1.9	1.2	n.d.	n.d.	34	1.8	1.4	5.8	24	18	15	96
Vernolate	n.d.	n.d.	n.d.	52	n.d.	n.d.	n.d.	4.2	n.d.	n.d.	n.d.	54

* Unresolved GLC peaks could not be measured.

Table XII. Major Pesticides Found in Evaporation Beds Sampled During 1981 and 1982. Levels Shown are Average Residues (>5.0 ppm) for a 0-12" Depth

Pesticide	West Side		Kearney		Lindcove		Tulelake		Davis	
	1981	1982	1981	1982	1981*	1982	1981	1982	1981	1982
Atrazine	n.d.	n.d.	<5.0	<5.0	<5.0	113	n.d.	n.d.	28	11
Carbaryl	<5.0	<5.0	n.d.	n.d.	167,600	<5.0	n.d.	n.d.	n.d.	n.d.
Chlorpyrifos	134	43	<5.0	<5.0	339	15	<5.0	<5.0	n.d.	n.d.
DACTHAL	243	206	6.3	5.4	n.d.	n.d.	n.d.	n.d.	n.d.	n.d.
DEVRINOL	81	74	27	13	**	1346	n.d.	<5.0	n.d.	n.d.
Endosulfan	<5.0	5.8	8.8	9.4	n.d.	n.d.	13.6	136	n.d.	n.d.
Parathion	n.d.	n.d.	n.d.	n.d.	56	508	7.6	30	n.d.	n.d.
Simazine	129	26	40	46	16,820	300	11.8	n.d.	7.9	2.3
Trifluralin	489	192	<5.0	22	n.d.	114	9.8	50	n.d.	<5.0

* The bed was ~90% flooded, permitting only one sample.

** GLC response was masked by other peaks, preventing quantitation.

very much, did not have average residues above 5 ppm. The one exception to moderate soil pesticide levels was the evaporation bed at Lindcove Field Station. These pesticide residues were very high and extremely nonuniform. In one sampling, for example, carbaryl residues ranged from below detectable levels in one part of the bed to 200,000 ppm in another section. Also, soil replicates with the highest residues tended to be in the corner of the evaporation bed closest to the entrance gate. Because of these findings, it is questionable whether this particular evaporation bed was being used as designed, or whether solid pesticide wastes were sometimes being dumped on the surface.

The rate of pesticide loss from a soil and water surface is dependent on numerous factors (5) including formulation type, presence of oils in the bed, the amount of water and the solubility of the pesticide in the aqueous-organic layer if and when the bed is flooded, and the level of organic matter that might support biological activity in the aerobic or anaerobic forms. In order to estimate the loss of pesticides over a one-year period, Table XII lists those pesticides considered to be of major importance in quantity and frequency of occurrence. In general, the pesticides did not appear to build up in significant quantities; even simazine levels had dropped considerably during a one-year period at the Lindcove Field Station. Since those five evaporation beds have been used extensively for many years the relatively low soil residues give evidence of their effectiveness. The high levels of carbaryl found in the Lindcove bed during 1981 followed by no detectable pesticides in 1982 may be attributed to the initial carbaryl deposit. Either the initial deposit was localized and subsequent sampling did not pick this up, or the pesticide was hydrolyzed by the lime in the bed. Most likely both of the above factors were responsible for its disappearance during this period. Some pesticides are somewhat seasonal in their use. For example, the cotton defoliant DEF is restricted to late summer and fall applications. One of the principal crops at the West Side Field Station is cotton, for which DEF is commonly employed for defoliation. It was in the beginning of the defoliation period when the late summer samples were taken in 1981 (Table VII). The station records had shown that DEF was not used in 1982 up to the time of our summer sampling. Average total soil residues in the 0-12 in. sampling dropped from 164 ppm in summer 1981 to 25.6 ppm by summer 1982. Levels of DEF could have been much higher by late fall of 1981 but this could not be determined without additional sampling and analyses. The data does show that residues are reduced appreciably for this one chemical during the course of a year when no new inputs are made.

The amount of soil residues found in the different soil depths, as shown in Tables VI to XI, indicated that a major trend for the pesticides was to concentrate in the upper 0-1 inch layer of soil. A theoretical explanation could be the "wick effect" reported by Hartley (6). As water evaporates off the soil

surface, water from below is drawn up by capillary action.
Pesticide compounds are then carried up to the surface by mass
transport. Under conditions of low humidity, water evaporation,
and hence transfer of pesticides to the surface, is increased. At
high pesticide concentrations, mobility could also be expected to
increase (7). One field station, Tulelake, appeared to be an
exception. In the Tulelake evaporation bed, pesticide residue
levels were higher in the 6-12 inch layer of soil. One explanation for this could be a leak in the bed's butyl rubber liner. If
the bed were leaking, some of the water's flow would be diverted
downward, and pesticide movement to the surface would be less. An
alternate explanation, which was reported later to us, was that
the bed was mixed with 6 inches of additional soil, and lime was
added just prior to our sampling. Which of these two factors is
responsible for unusual distribution at the Tulelake bed is not
known.

    The effect of incorporated lime on pesticide accumulation was
difficult to determine with absolute certainty from this study.
Reports that we had received showed that only three beds were not
limed. Two of the unlimed beds had another variable, which
involved storing the liquid wastes in 10,000 gallon storage tanks
until they were administered to the bed. The aqueous solution
coming from the tanks had a pH of 6.0. What this storage does to
the decay of pesticides before they reach the bed is uncertain.
It did appear, however, that with one exception (Lindcove) those
beds containing the highest quantity of lime and highest pH values
(Table XIII) generally had the lowest residues. South Coast Field
Station in particular, which was white with lime even at the 6-12
in. depths, also had the lowest residues. Lindcove Field Station,

Table XIII.   Average Soil pH from Evaporation Bed Samples Located
              at University of California Field Stations

		pH			
Field Station	0-1"	1-6"	6-12"	$\bar{x}$	S.D.
South Coast*	8.25±0.87	12.08±0.15	9.75±2.63	10.00	2.19
West Side	6.50±0.00	7.43±0.61	7.21±0.64	7.05	0.63
Kearney	6.50±0.00	6.50±0.00	6.50±0.00	6.50	0.00
Lindcove*	8.05±0.17	7.72±0.30	7.35±0.17	7.71	0.36
Deciduous Fruit*	8.40±2.48	12.02±0.15	9.02±2.09	9.82	2.37
Hopland*	7.00±0.58	10.1 ±2.37	6.83±0.58	8.08	2.10
U.C. Davis	7.62±1.18	6.20±0.00	6.20±0.00	6.68	0.93
Sierra Foothill*	7.25±0.50	11.50±2.0	8.23±2.92	9.01	2.65
Tulelake*	9.00±0.00	11.05±1.46	8.25±0.50	9.64	1.51

* Stations which had lime incorporated into their beds

on the other hand, had the highest residues, particularly at the surface even though lime had been added to it. Although the pH of the soils at Lindcove were above 7.3, none of the samples exceeded 8.0. All beds which had average pH values for the 0-12 in. depth greater than 9.0 had relatively low pesticide residue levels. As mentioned previously, the primary reason for low residue levels at Hopland and Sierra field stations was the infrequent use of the beds. The Deciduous Fruit Field Station was used moderately.

Due to the movement of the pesticides to the bed surface, air samples were taken to determine any volatilization and subsequent concentration in the air along the berm on the downwind side of the bed. In most instances, the top of the berm was only about 12 vertical inches above the bed surface. Spencer and Farmer (8) have reviewed the literature on the transfer of pesticides into the atmosphere. Even though pesticide volatility is related to vapor pressure of the chemical, there are many factors influencing the effective vapor pressure from soil and water surfaces.

Table XIV shows the levels of pesticides in air samples in the five field stations that had detectable levels. In general, the pesticide air levels found did compare favorably with the vapor pressure and residue levels of chemicals in the top 0-1 inch surface of soil. These same conclusions have been made from previous studies (9). However, levels were not always detected in the quantities that might be predicted (4), perhaps due to other variables such as inconsistent wetting of the beds, oil film in some of the beds and unidentified foreign matter on the surface of some beds. Although the beds were all made up of sandy loam, the degree of sand, silt and clay varied appreciably.

Table XIV. Pesticides in Air Samples Taken Over U.C Evaporation Beds During 1982

	West Side		Kearney		Lindcove		U.C. Davis		Tule-lake
Pesticide	4/29	9/17	4/28	9/16	4/28	9/16	5/12	10/14	9/8
Atrazine	n.d.	n.d.	n.d.	n.d.	n.d.	n.d.	0.17	0.11	n.d.
Chlorpyrifos	0.08	n.d.	n.d.	n.d.	n.d.	n.d.	n.d.	n.d.	n.d.
DEF	0.19	0.02	n.d.	n.d.	n.d.	n.d.	n.d.	n.d.	n.d.
Diazinon	n.d.	n.d.	n.d.	0.08	n.d.	0.36	n.d.	n.d.	n.d.
Parathion	n.d.	n.d.	n.d.	n.d.	0.04	5.87	n.d.	n.d.	0.02
Simazine	0.14	n.d.	n.d.	n.d.	0.05	n.d.	n.d.	0.01	n.d.
Trifluralin	1.48	n.d.	0.18	4.15	2.94	1.39	n.d.	n.d.	0.67

($\mu g/m^3$)

Summary and Conclusions

The evaporation beds used at University of California field
stations provided an economical method for on-site disposal of
dilute pesticide washings created by rinsing of used containers
and spray equipment. Large volumes of dilute pesticide solutions
were concentrated down to more manageable levels. When the evaporation beds were used as designed, and under the conditions of
this study, high pesticide residues did not tend to build up after
6 to 10 years of use. Pesticide residues did tend to concentrate
in the top 0-1 inch of soil, possibly due to mass transport as the
water moved toward and evaporated from the surface. Incorporating
lime into the soil of the bed also appeared to accelerate the
degradation of some pesticides. In order to maintain a reasonable
level of safety, evaporation beds should be monitored at least
once each year.

Literature Cited

1. Dillon, A.P. "Pesticide Disposal and Detoxification;
    Processes and Techniques"; Noyes Data Corporation: Park Ridge,
    NJ, 1981; Part II.
2. Greene, K., personal communication.
3. Wehner, T.A.; Woodrow, J.E.; Kim, Y.H.; Seiber, J.N., in
    "Identification and Analysis of Organic Pollutants in Air";
    Keith, L.H., Ed.; Butterworth Publishers: Woburn, MA, 1984;
    pp. 273-290.
4. Smith, R.T.; Atkinson, K. "Techniques in Pedology"; Urwin
    Bros. Ltd., The Greshorn Press, 1975.
5. Sanders, P.F.; Seiber, J.N. Chemosphere 1983, 12, 999-1012.
6. Hartley, G.S., in "Pesticidal Formulation Research"; Gould,
    R.F., Ed.; ADVANCES IN CHEMISTRY SERIES No. 86, American
    Chemical Society: Washington, D.C., 1969; pp. 115-134.
7. Davidson, J.M.; Rao, P.S.C.; Ou, L.T.; Wheeler, W.B.;
    Rothwell, D.F. "Adsorption, Movement, and Biological
    Degradation of Large Concentrations of Selected Pesticides in
    Soils"; EPA-600/2-80-124; Municipal Environmental Research
    Laboratory, U.S. Environmental Protection Agency: Cincinnati,
    OH, August 1980; pp. 15-42.
8. Spencer, W.F.; Farmer, W.J., in "Dynamics, Exposure and Hazard
    Assessment of Toxic Chemicals"; Haque, R., Ed.; Ann Arbor
    Science Publishers: Ann Arbor, MI, 1980; pp. 143-161.
9. Nash, R.G. J. Agric. Food Chem. 1983, 31, 210-217.

RECEIVED February 13, 1984

On-Site Pesticide Disposal at Chemical Control Centers

TERRY D. SPITTLER, JOHN B. BOURKE, PAUL B. BAKER[1], JAMES E. DEWEY[2], THOMAS K. DeRUE[3], and FRANK WINKLER[3]

Pesticide Residue Laboratory, New York State Agricultural Experimental Station, Cornell University, Geneva, NY 14456

In cooperation with the USDA-Soil Conservation Service, Chemical Control Centers have been installed on many small fruit farms in the 20-300 acre range. These facilities, consisting of a water source, catch basin, leach lines, and pesticide storage, help minimize danger to the worker and damage to the environment in the mixing and filling stages of pesticide spraying operations. In this study a series of surface water and deep soil samples were analyzed to detect any migration or runoff of waste pesticides from typical Chemical Control Centers. Entomological evaluation of soil biota and monitoring of dermal exposure to pesticides of mixer-applicators took place throughout the 1980 season. No adverse effects as a result of the Chemical Control Centers were detected.

On-site pesticide disposal is restricted to and intended for locations in or immediately adjacent to the agricultural areas being treated, and generally limited to dilute pesticide solutions from tank rinsing, equipment cleaning, overflow, and spillage in the mix-fill operation, and occasional disposal of excess tank mix. The concept is as old as the usage of agricultural chemicals, but regrettably, the site has frequently been the farm yard, barn floor cracks, or a convenient stream.

[1]Current address: Department of Entomology, NYSAES, Cornell University, Geneva, NY 14456

[2]Current address: Department of Entomology, Cornell University, Ithaca, NY 14850

[3]Current address: USDA-Soil Conservation Service, Wayne County Soil and Water Conservation District, Sodus, NY 14551

0097-6156/84/0259-0117$06.00/0
© 1984 American Chemical Society

Localization of pesticide storage, mixing, filling and
equipment cleaning has been practiced on large agricultural
operations for many years because the economics of scale justify
the maintenance of separate facilities. Small owner-operated
farms frequently cannot support this option; thus chemical
storage is relegated to any available corner, and most farmers,
out of convenience or necessity, fill their spraying equipment
near streams, drainage ditches, and farm ponds. Overflow from
mixing and filling operations and from tank rinsing is hazardous
and easily washed into nearby water sources. Eventually, the
desirability of restricting these operations so as to minimize
pesticide contamination was realized; as a result of this evolution, three distinct approaches came to be recognized; 1. that
which was legal; 2. that which was sensible; and 3. that which
was occurring as general practice.

For six years prior to this study, the Soil Conservation
Service and the Wayne County, NY, Soil and Water Conservation
District had been installing Chemical Control Centers (CCC) on
individual farms. The purpose of the Chemical Control Centers
is anti-pollution, providing an area on the farm where several
classes of agricultural chemicals, including insecticides, herbicides, and fungicides, may be handled in a safe manner,
thereby minimizing harmful effects upon the environment and the
farm workers during certain stages of the spraying operation.

Chemical Control Centers are designed and constructed according to standards and specifications developed by the Soil
Conservation Service. They also conform to all applicable local
and state ordinances. The basic components are a loading pad of
reinforced concrete and a leaching field. Other components may
include a water source, water lines, storage tank(s), electrical
service, and a pesticide storage building. Each center differs
in final form, usually due to site considerations and farmer
preference. The primary consideration in designing and locating
the components of the center is the availability of moderately-
to-well-drained soil for the leach lines.

The centers appeared to be doing the job for which they
were designed, and by 1980 they had become an accepted way of
dealing with chemical usage in an agricultural setting of
diverse operations on 20–300 acre tracts. Eventually, the New
York State Department of Environmental Conservation voiced
several concerns regarding their operation:

1. Was there any significant migration of chemicals away
   from the leach lines and into the surrounding soil,
   water supplies or to the surface?
2. Was there increased risk of exposure to pesticides by
   the agricultural workers as a result of the CCCs?
3. Were there any harmful effects upon the ecosystem in
   the vicinity of the centers and their leach lines?
4. Finally, should farms with these now identifiable centers be classified as point-source polluters, and, in
   recognition of this status, be required to obtain a

state license, file an environmental impact statement, submit a monthly report of the type and quantity of material discharged, and install automated monitoring equipment?

It was decided to address these questions simultaneously. Several cooperating farms in Wayne County were found with similar CCCs consisting of pad, leach lines, remote water source, water storage tank, and adjacent pesticide storage. A control farm was also located. Most pesticide operations were performed by the owner—operators or regular employees, and they agreed to personally cooperate.

CCC Locations

Farm A (Figure 1) has a CCC which had been in operation for five years. It is situated 86 feet away from a shallow pond bordering the orchard, and four feet upgrade from the water surface. The upper and lower leach lines extend 77 and 71 feet, respectively, between rows of apple trees parallel to the ponds edge. A buried tile located 20 feet north of the CCC helps drain the orchard hillside and also accommodates runoff from Lake Road. Water is pumped from an intake approximately two feet below the surface of the pond. This site is several miles removed from the main farm complex on Hilton gravelly loam.

The Chemical Control Center for Farm B, which was in its first year of operation, is an integral part of the farm compound. Located just at the edge of the building complex, pesticide storage is in the back portion of a service garage, and the concrete pad is also the refueling site for agricultural vehicles. Mix water is pumped 150 feet from the upper of two water impoundments that eventually form the swamp located in the northeast corner of the orchard, a swamp that also receives water from a drainage tile coursing west-northwest through the orchard. Leach lines from this facility border, but do not enter, the orchard, which is situated on Williamson silt loam.

An abandoned orchard (Farm C) which had not received any pesticide applications for three years was used as a control (Elnora loamy fine sand).

Sampling and Analysis

Water samples were taken on April 21, August 6, and December 11, 1980 as described and illustrated in Table I and Figure 1, respectively. All samples were analyzed for the following pesticides, using routine methods derived from the Pesticide Analytical Manual, PAM, (1), and shown to contain less than the indicated amounts (parts per million): captan <0.2; difolatan <0.1; carbaryl <0.01; methylparathion <0.1; permethrin <0.01; azinphosmethyl <0.1. Cited thresholds were above the detection limits.

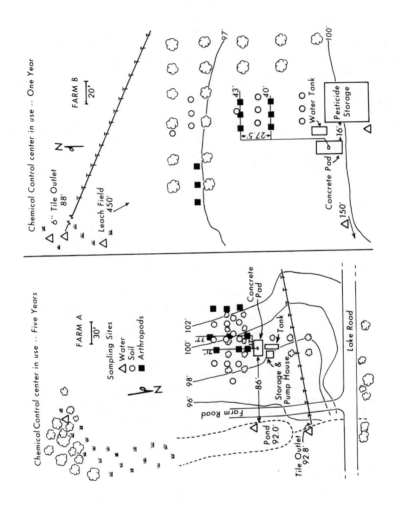

Figure 1.   Chemical Control Centers and Sampling Sites on Farms A and B.

Table I.   Water Sampling Sites for Farms A and B

FARM	DESCRIPTION	DATE
A	Swamp & pond source in woods	5/21/80
A	Swamp & pond source in woods	8/6/80
A	Pond below orchard and CCC	5/21/80
A	Pond below orchard and CCC	8/6/80
A	Orchard drainage tile outlet at pond	5/21/80
A	Orchard drainage tile outlet at pond	12/11/80
B	Pond, CCC and swamp water source	5/21/80
B	Pond, CCC and swamp water source	8/6/80
B	Deep well tap adjacent to CCC	5/21/80
B	Swamp below orchard, above drainage tile	5/21/80
B	Swamp below orchard, below drainage tile	8/6/80
B	Outlet of orchard drainage tile at swamp	5/21/80
B	Outlet of orchard drainage tile at swamp	12/11/80

The Chemical Control Centers and associated leach fields, being located in areas actively utilized in orchard operations, made actual excavation, measurement, and removal of sample sediment impossible without disrupting spray and cultivation schedules. Water tables were known to be considerably below the leach lines in the CCCs observed (8-20 feet), so that surface water sources down gradient from the areas were determined to be the most likely points to find any migration and contamination that might be occurring. No adjacent test wells into the water table were available. Soil samples to levels below the leach tiles were taken at several intervals progressing down gradient from the CCCs. No soil samples were taken any closer than 18 inches to a buried tile to avoid damage by the drilling and coring operations.

Shallow soil samples were taken with a soil probe on March 24, 1980, the two-inch diameter cores being divided into 0-3, 3-7, and 7-10 inch horizons. The top layer comprised of turf and roots was discarded as it would be contaminated by pesticides from tree spraying operations. The top layer was likewise discarded when deep soil samples were collected on August 7, 1980. A gasoline powered three-inch posthole auger was operated through a likesized hole in the bottom of a wooden box. Material brought to the surface from a specified depth was collected in the box, bagged, and the successive lower horizon was subsequently bored and collected. While some contamination of lower samples by soil in the upper horizons was inevitable, the method was potentially capable of identifying the sampling level at which a pesticide first appeared. Soil sampling sites are described in Table II (see Figure 1, also). Analyses using

Table II. Sample Sites for Soil Cores Taken at Farm A and Farm B

FARM	LOCATION	HORIZON (IN)	DATE
B	Up gradient from leach field (40')	0-3	4/24/80
B	Up gradient from leach field (40')	3-7	4/24/80
B	Up gradient from leach field (40')	7-10	4/24/80
B	In leach field	0-3	4/24/80
B	In leach field	3-7	4/24/80
B	In leach field	7-10	4/24/80
B	Down gradient from leach field (40')	0-3	4/24/80
B	Down gradient from leach field (40')	3-7	4/24/80
B	Down gradient from leach field (40')	7-10	4/24/80
A	Up gradient from leach field (40')	0-3	4/24/80
A	Up gradient from leach field (40')	3-7	4/24/80
A	Up gradient from leach field (40')	7-10	4/24/80
A	In leach field	0-3	4/24/80
A	In leach field	3-7	4/24/80
A	In leach field	7-10	4/24/80
A	Down gradient from leach field (40')	0-3	4/24/80
A	Down gradient from leach field (40')	3-7	4/24/80
A	Down gradient from leach field (40')	7-10	4/24/80
B	Edge of leach field near line	0-14	8/7/80
B	Edge of leach field near line	14-34	8/7/80
B	Edge of leach field near line	34-72	8/7/80
B	Down gradient from leach field (50')	0-18	8/7/80
B	Down gradient from leach field (50')	18-30	8/7/80
B	Down gradient from leach field (50')	30-72	8/7/80
A	Up gradient from leach field (40')	0-14	8/7/80
A	Up gradient from leach field (40')	14-36	8/7/80
A	Up gradient from leach field (40')	36-66	8/7/80
A	Edge of leach field near line	0-14	8/7/80
A	Edge of leach field near line	14-39	8/7/80
A	Edge of leach field near line	39-72	8/7/80
A	Down gradient from leach field (60')	0-14	8/7/80
A	Down gradient from leach field (60')	14-30	8/7/80
A	Down gradient from leach field (60')	38-60	8/7/80

PAM methods for azinphosmethyl and methylparathion -- two chemicals being frequently used in the subject orchards -- showed the former to be <0.1 ppm, and the latter to be <0.05 ppm for all soil samples.

Pesticide mixers and applicators at Farms A and B, and also at an active farm not having a CCC, were monitored for topical (dermal) exposure throughout the 1980 season. No increased exposure was measured as a result of using the Chemical Control Centers. A paper addressing potential applicator exposure in this and similar situations is in preparation(2).

Soil cores collected throughout the growing season (May-October) were monitored for arthropod populations as a measure of possible environmental repercussions of the CCCs. Soil cores, 5 x 3 in., were taken weekly from directly over the leach lines and uphill between the trees (check) at Farm A and Farm B (9 reps/site). A third orchard (Farm C), which had been abandoned for a least three years and had had no association with a CCC, was used as a control (3 reps/site). Berlese funnels were used to collect arthropods driven from the core samples into alcohol. Specimens were sorted to general categories, counted and either identified to family or sent out for taxonomical identification. The number of taxa found over all locations -- as determined from a similarity matrix -- is fairly consistent. Results were similar when all locations were polled to compare taxa found above the leach lines vs the checks and from Farm A vs Farm B. This would imply that there were no detrimental entomological effects from the pesticides present in the leach lines(3).

Conclusions

Pesticides from the CCCs were shown not to have leached into any of the available ground water sources. In addition, the soil core analyses indicated that disposed chemicals had traveled less than 18 inches from the leach lines in five years of operation. Within the defined time frame, the CCCs were confining the chemicals. A long-term follow-up survey, including test wells, would be useful to determine if migrating pesticides eventually reach the water table. The absence of any increased risk to applicators, or of any environmental drawbacks, as measured by soil arthropod populations, was established. Consequently, the New York State Department of Environmental Conservations has regarded these installations as normal agricultural operations and not required further documentation.

Literature Cited

1.   "Pesticide Analytical Manual" II; Food and Drug Administration: Washington, D.C., 1982.

2.   Spittler, T. D.; Bourke, J. B. in "Risk Determination for
     Agricultural Pesticide Workers from Dermal Exposure";
     Honeycutt, R. C., Ed.; ACS SYMPOSIUM SERIES, American
     Chemical Society; Washington, D.C., in preparation.
3.   Baker, P.B.; Hoebeke, E. R.; Barnard, J.; Spittler, T. D.
     Envir. Entomol., in press.

RECEIVED March 6, 1984

Treatment of Pesticide-Laden Wastewater by Recirculation Through Activated Carbon

EDMUND A. KOBYLINSKI[1], WILLIAM H. DENNIS, JR., and
ALAN B. ROSENCRANCE

U.S. Army Medical Bioengineering, Research and Development Laboratory, Ft. Detrick,
Frederick, MD 21701

The Carbolator 35B, a recirculatory carbon filtration system, successfully treated pesticide wastes
in both pilot-scale and field tests. In pilot-scale
tests, the system was challenged with 400 gallons of
water containing 20, 60 and 100 mg/L of each of
seven pesticides (baygon, dimethoate, diazinon,
ronnel, malathion, dursban, and 2,4-D). The pesticide waste was pumped through the Carbolator and
returned to the waste holding tank. The tank contents were analyzed by gas chromatography and thin
layer chromatography. A TLC method was developed to
perform pesticide analysis in the field. A simple
laboratory recirculating system using 4 liters of
waste was also built to simulate the Carbolator
system. A mathematical model was developed to predict the disappearance of the pesticides from the
waste holding tank.

Background

In the 19th century, various carbons were studied for their
ability to decolorize solutions and adsorb compounds from gases
and vapors. Commercial applications of activated carbon began
early in the 20th century. Solutions containing phenols, acetic
acid, herbicides, dyes, chlorophenols, cyanide and chromium have
been successfully treated by carbon adsorption ([1]).

Activated carbon is specially treated to give a high adsorption capacity. The adsorption capacity is dependent upon surface
area. Most of the available surface area lies within the internal
pores of the carbon. One gram of carbon can have a surface area
from 500–1500 m^2. Concentration gradients between the internal

[1]Current address: Atlantic Research Corporation, 5390 Cherokee Avenue, Alexandria,
VA 22312

pores of the carbon and the bulk liquid provide the driving force
for the diffusion of the contaminant into the carbon pores which
controls the overall rate of contaminant removal. The bulk liquid
is defined as that portion of fluid far enough away from the
particle surface to not have a concentration gradient. The rate of
diffusion within the pore is set by the type of carbon, geometry of
the pore and degree of activation. Once a type of carbon is chosen,
only the rate of diffusion from the bulk liquid to the liquid
boundary surrounding the carbon prior to entering the pores can be
controlled. The liquid boundary layer is defined as that section
of liquid that has a concentration gradient. The thickness of the
boundary layer is affected by turbulence and is compressed during
periods of high turbulence.

Several methods can be used to expose a liquid waste to
activated carbon. The batch slurry method is the simplest.
Activated carbon is added to a batch volume of liquid waste and is
allowed to stand quiescently. In this case, the boundary layer
stretches from the carbon surface at the bottom of the tank to the
liquid surface resulting in a very slow diffusion rate. The
diffusion rate can be greatly accelerated by agitation. Mixing
will compress the boundary layer and increase the concentration
driving force. The carbon can be separated by decanting the
liquid. This method requires a long holding time but is not
manpower intensive.

Downflow through packed columns is another contacting process. This type of contact can handle high flow rates but the waste
stream cannot contain large quantities of suspended materials. The
suspended solids will clog the column and induce a high pressure
drop through the column. This process requires frequent back-
washing.

Upflow expanded bed carbon columns are a third contacting
process. The liquid waste enters the bottom of the carbon column
and expands the carbon bed. Higher suspended solids concentrations
can be handled in this system because of the greater distance
between particles in an expanded bed.

Both the upflow and downflow carbon adsorption systems can be
operated in a once through mode or in a recirculation mode. In a
once through mode, the carbon column effluent must be continuously monitored for breakthrough of the contaminant from the
carbon bed. This requires considerable lab support. The monitoring
requirements for a recirculation system are considerably less than
for a once through system because of the batch treatment aspects in
a recirculating system. The recirculating system will make less
efficient use of the carbon charge. The carbon charge in a
recirculatory system cannot be allowed to become saturated because
of the low concentration driving force and a lack of available
adsorption sites will tend to greatly increase the treatment time.
In a once through system, the initial sections or columns of carbon
continually see a high pesticide concentration and can become
saturated without hurting the overall system efficiency. As each

column is saturated, it can be removed from service. The next column in the series will then become the first column. Replacement carbon columns would be placed at the end of the column series. Each column in a once through system can be run until saturated, thereby making complete use of the carbon charge.

Introduction

The disposal of pesticide-laden wastewater produced by small generators, which include commercial and government pesticide applicators, is a current problem. According to Public Law 89-272 (Resource Conservation and Recovery Act), the responsibility for the proper disposal of hazardous wastes lies with the waste generator. Good management of the quantity of pesticide solutions prepared will result in small waste volumes, however, the bulk of the waste produced will come from the cleaning of pesticide application equipment and from the required rinsing of pesticide containers, especially 55-gallon drums. Carbon adsorption offers a simple and inexpensive solution to clean-up of these waste pesticides.

Recognizing the potential health and pollution problems resulting from the operation of a pest control facility, the U.S. Army Training and Doctrine Command (TRADOC) designed and built a pest control facility at Ft. Eustis, Virginia, to meet all current and anticipated health and environmental regulations regarding such facilities. A common drainage system links the pesticide mixing room, storage room and outdoor washdown area to a sump. The outdoor washdown area was designed in such a way so as to minimize the amount of rainwater entering the drainage system. The liquid waste collected in the sump is pumped into a storage tank for either treatment or disposal.

Prior to construction of the Ft. Eustis pest control facility, TRADOC requested the U.S. Army Medical Bioengineering Research and Development Laboratory (USAMBRDL) to investigate the feasibility of using an activated carbon filtration/absorption system to treat the pesticide-laden wastewater. The final system design was to: 1) not be manpower intensive; 2) not require extensive analytical monitoring, and 3) be compatible with both the types and quantities of pesticide wastes produced by a typical pest control facilty. The initial research was funded jointly by EPA and TRADOC.

An inexpensive commercial carbon filtration system (Carbolator) was found to have considerable potential in solving the waste treatment problem at pest control facilities. The Carbolator operates by the recirculation of wastewater through a bed of activated carbon.

Materials and Methods

Synthetic Wastewater. Between 1977 and 1981, samples of wastewaters generated at the pest control facilities at Ft. Eustis, Va. and Ft. Knox, Ky., were received and analyzed to determine the composition of typical wastes. In addition, contacts with pest control supervisors were made to determine the types and quantities of pesticides employed at these sites. From this information, the types of pesticides that would be used in the laboratory testing of the Carbolator system were set. Pesticides chosen for the Carbolator tests were chloropyrifos (Dursban), 2,4-D-LVE (low volatile ester), Diazinon, Dimethoate (Cygon), Fenchlorphos (Ronnel), Malathion and Propoxur (Baygon). Structures for these pesticides are presented in Figure 1.

Gas Chromatography Analysis of Water for Pesticides. All analyses for pesticides in water were done by gas chromatography. Solvents used for extraction were checked by gas chromatography for purity and interferences and all glassware used in the extraction was cleaned in a chromic acid/sulfuric acid mixture. Standards consisted of mixtures of various pesticides (actual commercial formulations) suspended or dissolved in water. These aqueous standards were extracted in the same manner as unknown solutions. The standard concentrations encompassed the concentration of unknowns to be determined. A standard curve normally consisted of a set of four pesticide concentrations. Blanks were run and an internal standard (eicosane) was used. The internal standard concentration was kept constant for all analyses. The conditions for GC analysis were guided by the pesticides expected in the water. For the more complex mixtures, such as those employed in the synthetic waste and those encountered in the field, a 6 ft., 3 percent SE-30 on GAS CHROM Q column sufficed. A typical chromatogram of a complex pesticide mixture is shown in Figure 2. (2)

Figure 3 shows a flow chart that outlines the protocol for analysis both of synthetic wastewater and waters encountered in field tests at Ft. Eustis, Va.

Analysis of On-Site Wastewaters for Pesticides by Thin-Layer Chromatography. A field method to identify qualitatively and semi-quantitatively the pesticide constituents of a pesticide-laden wastewater was developed. The field method was developed using thin-layer chromatography (TLC). TLC gives a presumptive test for the presence of specific pesticides and within 30 minutes an estimate of their concentrations. TLC may also reveal the presence of unknown substances. The field application of thin-layer chromatography requires a skilled chemist, but no expensive equipment. The following protocol describes the on-site use of TLC.

Twice extract a 200 mL aliquot of the wastewater with 25 mL of CH_2Cl_2 in a 250 mL separatory funnel. Combine both CH_2Cl_2 extracts and add a few grams of anhydrous Na_2SO_4 to absorb moisture. Pour a portion of the CH_2Cl_2 extract into a 20 mL glass vial and allow the CH_2Cl_2 to evaporate in the open air. As the CH_2Cl_2 evaporates, add more of the extract until the entire CH_2Cl_2 extract has been

COMMON NAME	ALTERNATE NAME	CHEMICAL NAME	STRUCTURE
Chlorpyrifos	Dursban	0,0-diethyl-0-(3,5,6-trichloro-2-pyridyl) phosphorothioate	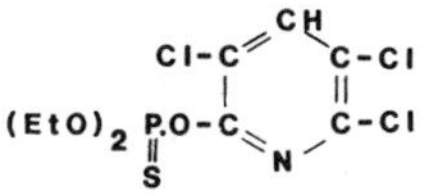
2,4-D-LVE	------	isooctylester of 2,4-dichlorophenoxy-acetic acid	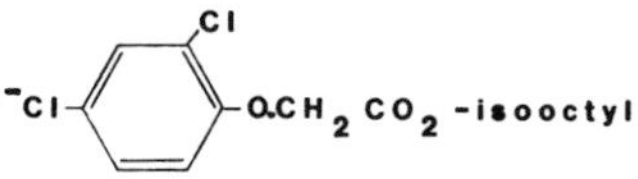
Diazinon	------	0,0-diethyl-0-(2-isopropyl-6-methyl-4-pyrimidinyl) phosphorothioate	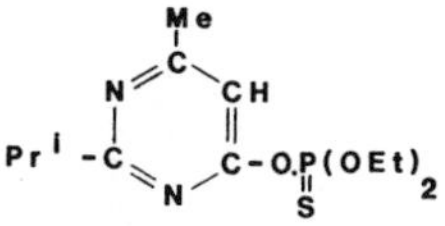
Dimethoate	Cygon	0,0-dimethyl-S-(N-methylcarbamoylmethyl) phosphorodithioate	$(MeO)_2 PS.S.CH_2 CO.NHMe$
Fenchlorphos	Ronnel	2,4,5-trichlorophenyl phosphorothioate	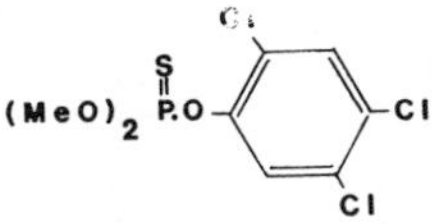
Malathion	------	dimethylmercapto-succinate S-ester of 0,0-dimethylphosphoro-dithionate	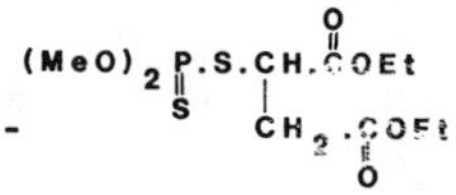
Propoxur	Baygon	2-isopropoxy-phenyl N-methyl carbamate	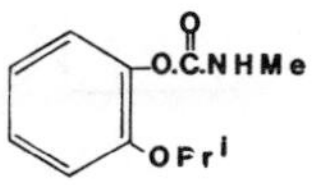

Figure 1. Synthetic waste components.

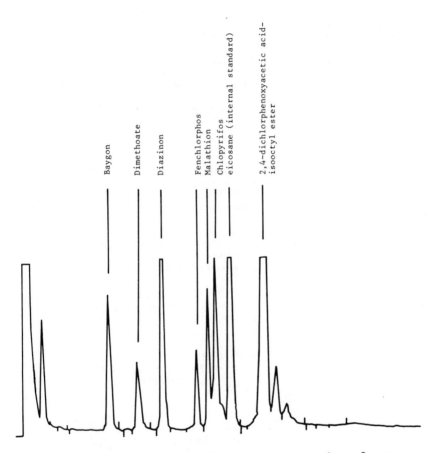

Figure 2.   Gas chromatogram showing the separation of seven pesticides. Reproduced with permission from Ref. 2. Copyright 1983, Marcel Dekker, Inc.

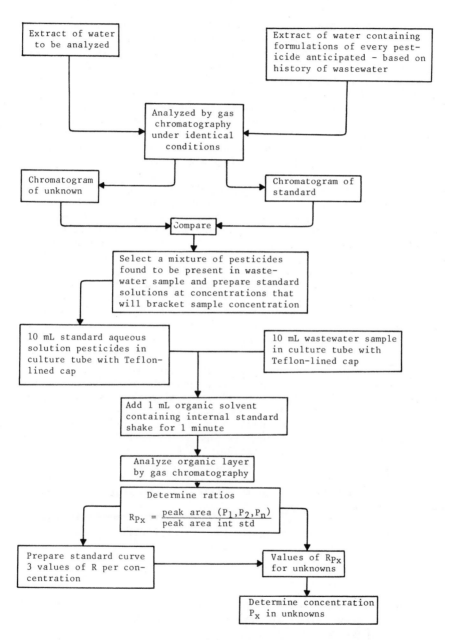

Figure 3. Flow chart showing the qualitative and quantitative analysis of a complex pesticide waste (<u>3</u>).

reduced to an oily residue. This residue will contain the pesticides that were suspended or dissolved in the wastewater. A heat gun (hair dryer) may be used to hasten evaporation of the CH_2Cl_2, but great care must be used to prevent loss of the extract by too vigorous boiling. Prior to analysis by thin-layer chromatography (TLC), the residue in the vial is reconstituted with 1 mL of fresh CH_2Cl_2. Ten μL of this solution is placed in a small spot, 1 cm above the bottom of a TLC plate (E. Merck, Silica gel 60, GF-254). At a point of 1-2 cm to the right and left of this spot, place 10 μL of a solution containing several pesticides of known concentration. The TLC plate is then placed into a developing jar containing a suitable solvent. When the solvent has risen about 12 cm by capillary movement, the TLC plate is removed from the jar, air dried, and observed under UV light. Since the silica gel contains a fluorescent substance, any UV-absorbing substances will show up as dark spots. Any observed dark spots are outlined in pencil. Finally, the TLC plate is sprayed with an acetone solution of TCQ (N,2,6-trichloro-p-benzoquinoneimine) and then heated with a heat gun until colored spots appear. Spraying with TCQ and the heating of the TLC plate should take place in the open air because of the irritancy of this reagent.

The spots present in the wastewater extract are matched with those of the known mixture. This comparison gives qualitative information regarding the wastewater extract. The relative size of matching spots gives a semi-quantitative estimate of the concentration of pesticide in the extract (3).

Carbolator 35B Filtration System. The Carbolator, a commercial unit produced by Sethco Corporation, operates by recirculating a stream of wastewater upward through a bed of carbon. After passing through the carbon bed, the wastewater is returned to the waste holding tank. The carbon bed is contained within two porous polypropylene bags. Both bags fit inside an 18-gallon reinforced epoxy tank with an o-ring seal lid. A plastic flow distributor rests beneath under the bags of carbon to reduce channeling. The pump for the system can be mounted externally on a steel mounting bracket. A schematic of the Carbolator recirculatory system is shown in Figure 4.

A mass balance, Equation 1, can be written around the waste streams leaving and entering the waste holding tank.

$$\frac{dX}{dt} = Cq_{out\ of\ tank} - (1-k)Cq_{returning\ to\ tank} \qquad (1)$$

Equation 1 describes the change of mass, pesticides, within the tank at any given time. The symbols used in this derivation are defined as follows:

V   =   the liquid volume in the waste holding
          tank in liters

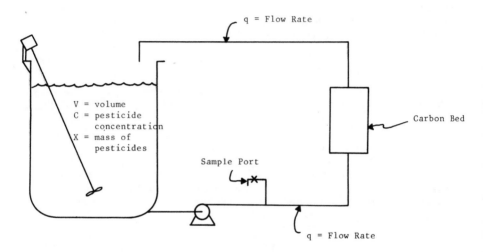

Figure 4. Carbolator process schematic.

q   =   volumetric flow rate through the Carbolator in liters per minute (LPM)

t   =   time in minutes

C   =   the pesticide concentration in the waste holding at any time in mg/L

C_O =   the initial pesticide concentration in the waste holding tank in mg/L

k   =   a pesticide removal efficiency factor through the Carbolator

X   =   the mass of pesticides in the waste holding tank any time in mg

The k term is required because the pesticide concentration leaving the Carbolator will not be zero, therefore (1-k) describes the pesticides remaining in the waste stream. Since X is really the concentration (C) within the tank times the tank volume (V), CV can be substituted for X as shown in Equation 2.

$$\frac{VDC}{dt} = kCq \qquad\qquad (2)$$

Rearranging the terms yields Equation 3 which can be integrated.

$$\frac{dC}{C} = \frac{kq}{V} dt \qquad\qquad (3)$$

Integration of Equation 3 results in a first order rate decay equation. Equation 4 describes the pesticide concentration in the wastewater which will decrease exponentially, but never reach zero.

$$C = C_o \exp(\frac{-kqt}{V}) \qquad\qquad (4)$$

Equation 5 shows that a semi-log plot of log $\frac{C}{C_o}$ vs. t will have a slope equal to $\frac{-kq}{2.303\ V}$

$$\log \frac{C}{C_o} = \frac{-kq}{2.303\ V} t \qquad\qquad (5)$$

This derivation assumes that the holding tank is completely mixed. When non-complete mixing conditions occur, the required treatment time will be longer than the time predicted by Equation 4.

Results and Discussion

    Pilot Scale Tests.   Three pilot-scale tests were made at USAMBRDL prior to the Ft. Eustis field tests. Measured amounts of the commercial formulations of the pesticides in the synthetic

waste (all were emulsifiable concentrates except for Baygon which
was a wettable powder) were separately added to 400 gallons of tap
water held in a 500 gallon tank. The concentrations of each of the
active ingredients were 20 mg/L for test 1, 60 mg/L for test 2, and
100 mg/L for test 3. A 1/4 hp gear-driven mixer with two 9 inch
diameter marine impellers provided the mixing. The wastewater for
all three tests was pumped through the Carbolator at 6.4 gpm and
returned to the holding tank. Ten mL aliquots were taken from the
wastewater tank periodically. These samples represented the
pesticide concentration in the holding tank at a given time. All
samples were analyzed by gas chromatography.

Table I presents the change in concentration of each pesticide
as a function of time at the three initial levels of 20 mg/L, 60
mg/L and 100 mg/L. The theoretical pesticide concentration is also
shown in Table I as calculated from Equation 4 with k = 1, which
assumes complete pesticide removal in one pass. The data are
graphed and presented in Figure 5 as tank concentration versus
time.

From these data, the efficiency factor, k, can be determined
with q = 6.4 gpm and V = 400 gallons. The dashed line in these
figures represents the decrease in pesticide concentration if k =
1. Values of k were determined from a least squares analysis of the
data. These k values are presented in Table II.

The slope of the lines presented in Figure 5 is defined as
k(q/v). The q/v term defines the turnover of the tank contents or
what is commonly referred to as the retention time. When q is
increased, the liquid contacts the carbon more often and the
removal of pesticides should increase, however, the efficiency
term, k, can be a function of q. As the waste flow rate is
increased, the fluid velocity around each carbon particle increases, thereby increasing system turbulence and compressing the
liquid boundary layer. The residence time within the carbon bed is
also decreased at higher liquid flow rates, which will reduce the
time available for the pesticides to diffuse from the bulk liquid
into the liquid boundary layer and into the carbon pores. From
inspection of Table II, the pesticide concentration also effects
the efficiency factor. k can only be determined experimentally and
is valid only for the equipment and conditions tested.

Bench Scale Tests. It was also sometimes desirable to evaluate the
effectiveness of the Carbolator concept on a solution containing
one or more pesticides without resorting to the full-scale (400
gallons of wastewater) treatment system. In order to simulate the
recirculation of a pesticide-laden wastewater through a bed of
carbon, the bench-scale apparatus shown in Figure 6 was assembled.
By treating a water containing several pesticides simultaneously
in this apparatus, it is possible to determine their relative rates
of adsorption to carbon. With this apparatus, it is also possible
to evaluate various types of granular carbon.

Measured amounts of various pesticide formulations are poured

Table I. Pesticide Concentration in 400 Gallons of Water During Treatment With Carbolator 35B

Time (min)	Theoretical[a]	Baygon	Dimethoate	Diazinon	Ronnel	Malathion	Dursban	2,4-D (LVE)[d]
0	20	20	20	20	20	20	20	20
30	12.3	-[b]	11.0	16.5	14.7	10.1	14.5	14.5
60	7.6	6.6	7.4	11.3	10.9	7.4	11.0	11.5
120	3.0	4.8	4.7	5.7	4.6	4.0	4.5	5.0
360	0.06	ND[c]	ND	1.1	0.6	1.7	0.7	0.7
840	10^{-4}	ND	ND	0.2	0.1	1.0	0.03	0.05
0	60	60	60	60	60	60	60	60
30	37	26	39	43	45	39	43	44
60	23	22	28	35	34	35	36	35
120	9	13	16	23	24	17	23	24
340	0.3	4	4	5.7	5.7	7	5	6
810		ND	ND	1.1	0.4	0.6	0.4	0.7
0	100	100	100	100	100	100	100	100
30	62	79	78	75	84	73	72	69
60	39	65	62	54	47	54	59	69
120	15	42	40	46	55	44	52	54
360	0.3	15	16	24	39	19	32	33
1250		0.5	ND	1.4	2	0.9	2.6	5.6

a. Based on $\frac{C}{Co} = e\,\frac{-Qt}{V}$ where q = 6.4 gallons/minute and V = 400 gallons.

b. Not analyzed.

c. Not detectable by GC

d. LVE – low volatile ester

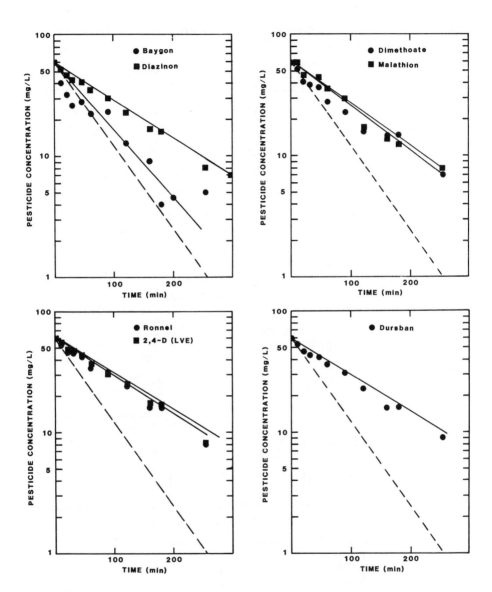

Figure 5. Carbolator pilot test 2 pesticide removal curves. Reproduced with permission from Ref. 2. Copyright 1983, Marcel Dekker, Inc.

Table II. Efficiency Factors (k) Calculated from Carbolator
           Pilot Tests

	Initial Pesticide Concentration, mg/L		
Pesticide	20	60	100
Baygon	0.83	0.81	0.32
Dimethoate	0.69	0.47	0.33
Diazinon	0.53	0.44	0.26
Ronnel	0.64	0.43	0.20
Malathion	0.61	0.43	0.29
Dursban	0.62	0.43	0.40
2,4-D (LVE)	0.56	0.40	0.16

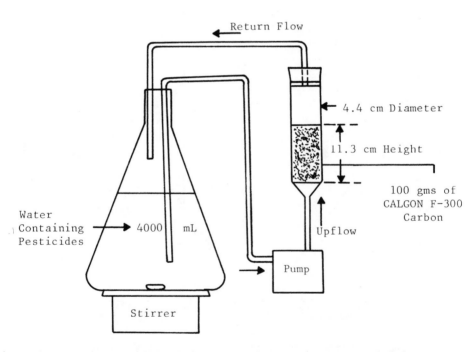

Figure 6.   Bench-scale carbon recirculatory system (3).

into a 5 L Erlenmeyer flask containing 4 L of tap water at room temperature. While this mixture is being stirred, place the wetted carbon (100 g dry weight) into the glass column. Prior to use, the carbon was soaked in water for 24 hours. Glass wool or a frit can be used to retain the carbon bed. Clean water (about 500 mL) is passed upward through the carbon bed (back washing) to remove carbon fines. The water is then drained from the carbon column and a connection is made between the stirred reservoir and the bottom of the carbon column through a peristaltic pump. A one-hole neoprene stopper is placed in the top of the glass column. A rubber hose carrys the wastewater from the top of the column back to the 5 L continuously mixed flask. The pump is turned on and adjusted to the desired flow, 300 mL/min. When the water has passed the top of the carbon bed, the carbon column is tapped to dislodge air pockets. When the water begins to spill back into the reservoir, a timer is started; this is t_0. Prior to starting the pump, an aliquot (3 mL) of the test water is removed from the flask. This sample is used to prepare a standard curve from which will be derived the concentration of pesticides in the reservoir. At various times during this recirculation, 1 mL, 5 mL, or 10 mL aliquots are removed from the reservoir and placed into screw cap (teflon-lined caps) tubes containing 1 mL of methylene chloride (CH_2Cl_2) that is spiked with 20 ppm of eicosane (internal standard). The aqueous sample was extracted with the 1 mL of CH_2Cl_2 and this extract was analyzed by gas chromatography (FID). Table III presents the changes in pesticide concentration for a bench-scale test run. The column was charged with 100 g of Calgon F-300 granular activated carbon. The data are presented graphically in Figure 7. It can be seen that 100 grams of carbon was able to remove the pesticide concentrations to less than 1 mg/L for 2,4-D and to below detection limits for the other six pesticides within 8 hours under the given conditions. Furthermore, both baygon and dimethoate closely follow the theoretical exponential curve indicating that these substances must be absent in the effluent from the carbon column after one pass.

Efficiency factors for the bench scale tests have been calculated and are presented in Table IV. The k values in Table IV are generally higher than the values presented in Table II for the pilot test at 100 mg/L of each pesticide. The bench-scale and pilot-scale systems were loaded as follows: 1) the q/v ratios were 0.016 min^{-1} for the pilot tests and 0.075 min^{-1} for the bench-scale tests and 2) the ratio of carbon to gallons of waste (which relates to the mass of pesticides at a given equivalent concentration) was 0.208 lb. carbon/gallon of water for the bench-scale tests and 0.1125 lb. carbon/ gallon of waste for the pilot tests. The fluid velocity through the open column cross-sectional area is 19.7 cm/min for the bench-scale tests and 8.3 cm/min for the pilot tests. This shows that the system loading has a large effect on the removal efficiency. If the bench-scale and pilot-scale system were operated under the same loading conditions, then the efficiency factors would be comparable to the actual Carbolator system.

Table III. Change in Concentration of Seven Pesticides During Small-Scale Simulation of the Carbolator Water Treatment Concept

Time (min)	Theoretical	Baygon	Dimethoate	Diazinon	Ronnel	Malathion	Dursban	2,4-D (LVE)
				Concentration in mg/L				
0	100	100	100	100	100	100	100	100
10	47	50	51	77	72	75	84	79
20	22	26	27	64	64	56	72	73
30	11	18	15	58	54	44	72	66
40	5	3	10	48	49	34	64	59
60	1	ND	4	25	32	17	47	43
120	0.01	ND	1	5	11	2	23	21
180		ND	ND	0.6	4	ND	10	11
360		ND	ND	0.2	0.5	ND	1	3
465		ND	ND	ND	ND	ND	ND	ND

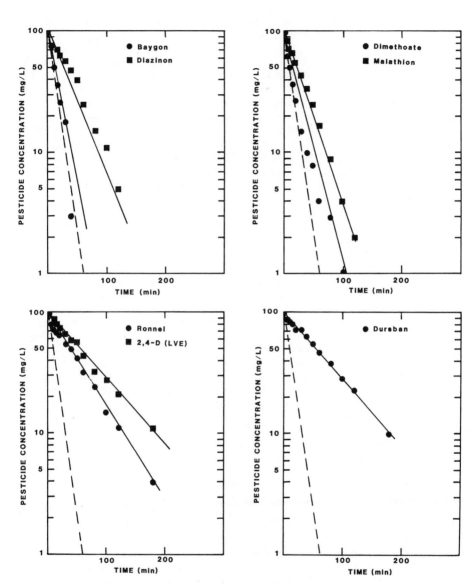

Figure 7. Pesticide removal curves for the bench-scale recirculatory carbon absorption system.

Field Tests.   A preliminary test with the Ft. Eustis wastewater indicated that an in-line filter was necessary for the removal of suspended solids from the water before the waste could be passed through the Carbolator. A drawing of the assembled apparatus is shown in Figure 8. As shown, water is taken from the bottom of the holding tank by a pump, P_1. The majority of the water passes back into the tank through line L_1. This keeps the contents of the tank constantly but not completely mixed. Part of the output of P_1 passes through a second pump, P_2, and then through a 20-inch particulate filter cartridge (F) in a polycarbonate housing. (The cartridge filter removes suspended solids, but not the dissolved pesticides.) The filtered water then passes into the bottom of the Carbolator with granular carbon (Calgon F-300) held in two porous polypropylene bags. The carbon-filtered water is returned to the tank through a flexible line, L_2. Water samples are taken before and after the cartridge filter and at the Carbolator oulet (L_2). All water samples were returned to the laboratory and analyzed by gas chromatography.

The first field test of the Carbolator treatment system was made in July 1981. This sytem (Figure 8) was assembled and connected to the wastewater tank containing about 600 gallons of water. On-site analysis of the waste by TLC showed the presence of malathion, dimethoate, and baygon with traces of diazinon and dursban. The wastewater was passed through the system (30 lb Calgon F-300) and returned to the waste tank. As the cartridge filters (2-micron size) became fouled (and the flow diminished) they were replaced. During the first 9 hours of operation, the flow through the Carbolator was monitored. The cumulative volume of water passing through the CARBOLATOR was determined by graphic integration of a flow versus time plot. The average flow was 6.1 gpm during the test. The system was allowed to operate unattended overnight and no problems were encountered. After 24 hours of treatment, no pesticides could be detected in the water by the on-site TLC analysis.

Gas chromatographic analysis of samples taken from the waste tank showed a steady decline in all three pesticides from the water. The results of the GC analysis of the water as a function of time are shown in Table V and graphed in Figure 9. The efficiency factors were calculated and are presented in Table VI.

The first field test was successful. Both dimethoate and malathion declined exponentially (Figure 9) and exhibited efficiency factors comparable to the efficiency factors found in the pilot tests. An efficiency factor could be calculated for baygon, but not for diazinon. Some sediment was present into the bottom of the holding tank which could have been slowly releasing baygon and diazinon in the bulk liquid. Nevertheless, after 24 hours of treatment, all pesticides were below the limit of detection.

A second field test was conducted in September, 1981. The carbolator system was assembled identical to that shown in Figure 8. In this trial, 410 gallons of wastewater (20°C) were processed

Table IV. Efficiency Factors Calculated From Bench-Scale Tests

Pesticide	k
Baygon	0.79
Dimethoate	0.58
Diazinon	0.36
Ronnel	0.23
Malathion	0.43
Dursban	0.17
2,4-D (LVE)	0.16

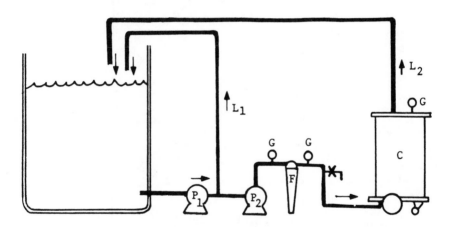

Figure 8. Carbolator field test process schematic. Reproduced with permission from Ref. 2. Copyright 1983, Marcel Dekker, Inc.

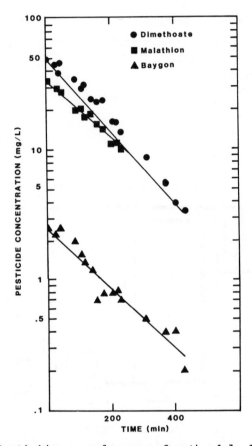

Figure 9. Pesticide removal curves for the July 1981 carbolator field test.

Table V. Pesticide Concentration in Ft. Eustis Wastewater as Carbolator Treatment
Progressed. Samples Taken Between Pump and In-Line Filter – July 1981
Field Test.

Time (min) In Operation	Pesticide Concentration in mg/L			
	Dimethoate	Malathion	Baygon	Diazinon
0	48.6	33.5	2.5	–
30	43.6	29.4	2.2	1.0
90	33.6	19.8	2.0	0.5
120	30.7	17.7	1.4	0.5
180	23.4	14.6	0.8	0.5
220	16.1	11.1	0.8	0.5
315	8.8	1.9	0.5	0.6
375	5.4	1.0	0.4	0.4
435	3.4	ND	0.2	ND
1440	ND	ND	ND	ND

ND = Not detectable.

Table VI. Efficiency Factors Calculated for the July 1981
          Carbolator Field Test

Pesticide	k
Dimethoate	0.62
Malathion	0.50
Baygon	0.51
Diazinon	0.51

q = 6.1 gpm
V = 600 gallons

with the Carbolator system (40 lb Calgon F-300 were used). The
system was operated over a 3-day period for a total of 20 hours. In
this test, the cartridge filter was charged with a 10 micron
cotton-wound filter element. The system flow rate was monitored.
The average flow rate from graphical integration was shown to be
5.82 gpm. Again, on-site TLC analysis was used to determine the
composition of the waste and the samples were collected between the
pump and the in-line filter and returned to the laboratory for GC
analysis. The results of GC analysis of the water as a function of
time are shown in Table VII and graphed in Figure 10. Efficiency
factors were calcualted and are presented in Table VIII.

The second field test was also successful. The plots in Figure
10 show a definite exponential decrease in pesticide concentrations. The efficiency factors for the field test shown in Table
VIII were lower than the expected efficiency factors resulting from
the pilot tests at comparable pesticide concentrations. This probably resulted from incomplete mixing within the waste holding
tank. The apparent rapid loss of malathion (more rapid than
theoretical) arose from the inhomogeneity of the waste. It was
observed during the first few minutes of operation that a heavy oil
had accumulated in the cartridge filter housing. This oil was apparently undissolved malathion that had pooled at the bottom of the
holding tank. It was learned subsequently that on the day before
the test began, an "empty" 55-gallon drum of 95% malathion was
rinsed into the sump. The immiscible malathion was transferred to
the holding tank by the sump pump. The malathion was pumped
directly into the carbon bed (where it was retained) at the start
of the operation. The aqueous phase of the waste contained very
little dissolved malathion. At the termination of this test, the
processed water showed no detectable malathion, baygon, dimethoate
or diazinon and the levels of dursban and of 2,4-D ester were 0.3
mg/L and 0.1 mg/L, respectively. Continued operation beyond 20
hours would have undoubtedly lowered these levels to their limits
of detection.

In both of these field tests, the application of thin-layer
chromatography was essential for characterizing the pesticides in
the water and in monitoring the progress of their removal by the
carbon filtration system.

Conclusions

The Carbolator carbon filtration system offers an economical
and dependable means by which small waste generators can effectively treat their pesticide-laden wastewater. The system is
portable and can be assembled from commercial items with a capital
cost of about $3,000 (1981 costs). The system was simple to operate
and with the aide of the thin-layer chromtography, the system was
easy to monitor.

Table VII. Pesticide Concentration in Ft. Eustis Wastewater as Carbolator Treatment Progressed.
Samples Taken Between Pump and In-Line Filter – September 1981 Field Test.

Time (min) In Operation	Concentration of Pesticides (mg/L)					
	Malathion	Baygon	Dursban	Dimethoate	Diazinon	2,4-D (LVE)
0	81.7	40.8	18.7	17.4	10.1	6.4
9	8.5	37.3	16.2	15.5	7.7	5.9
55	6.7	30.7	14.9	13.3	7.2	5.3
85	ND	24.6	11.8	9.9	5.6	3.8
140	ND	19.8	11.2	7.8	4.5	3.7
180	ND	13.8	8.1	5.9	4.1	3.5
240	ND	8.5	5.8	4.0	2.6	2.5
360	ND	3.3	3.4	1.7	1.4	1.3
480	ND	1.2	2.9	0.6	0.5	1.1
720	ND	0.3	0.9	ND	0.02	0.1
1200	ND	ND	0.3	ND	ND	0.1

ND = Not detectable.

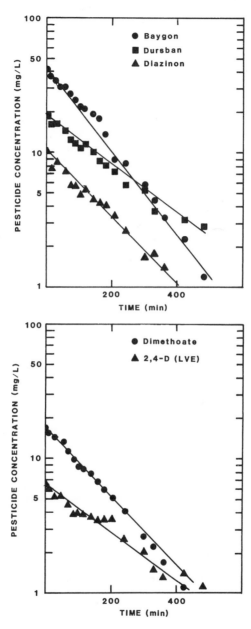

Figure 10.   Pesticide removal curves for the September 1981 carbolator field test (2).

Table VIII. Efficiency Factors Calculated for the September 1981
            Carbolator Field Test

Pesticide	k
Dimethoate	0.41
Malathion	3.47
Baygon	0.50
Diazinon	0.40
Dursban	0.29
2,4-D LVE	0.29

g = 5.82 gpm
V = 410 gallons

Literature Cited

1. DeRenzo, D.J., "Unit Operations for Treatment of Hazardous Industrial Wastes," Noyes Data Corporation, 1978.
2. Dennis, W.J. Jr. and Kobylinski, E.A., J. Environ. Sci. Health, 1983, B18(3), 317-331.
3. Dennis, William H., Jr.; Wade, Clarence W.R.; Kobylinski, Edmund A., and Rosencrance, Alan B., "Treatment of Pesticide-Laden Wastewaters from Army Pest Control Facilities by Activated Carbon Filtration Using the Carbolator Treatment System," Technical Report 8203, U.S. Army Medical Bioengineering Research and Development Laboratory, Frederick, MD., August 1983.

RECEIVED February 13, 1984

Treating Pesticide-Contaminated Wastewater
Development and Evaluation of a System

JOHN C. NYE

Purdue University, West Lafayette, IN 47907

Over the past five years, a system for removing
pesticides from the wash water produced by pesticide
applicators as they clean their equipment has been
developed. The system incorporates a two-stage
treatment process. The first step is the
flocculation/coagulation and sedimentation of the
pesticide contaminated wash water. The supernatant
from the first step is then passed through activated
carbon columns. This paper describes the
development of the system, the evaluation of the
system's adequacy to handle a wide variety of
pesticides, and the recommendations on the
implementation of this system to commercial
pesticide applicators.

Commercial pesticide applicators are faced with a serious problem
in the proper disposal of the large volumes of pesticide
contaminated wastewater that are produced during the cleanup of
application equipment. Various studies (Whittaker et al. 1982)
have reported that the typical agricultural pesticide applicator
will produce between 100 and 400 liters of pesticide-contaminated
wash water each time he cleans the equipment. For a typical
applicator, this amounts to approximately 20,000 liters of waste
annually from each piece of equipment (i.e., airplane or truck)
that he uses.

Very few techniques are available for pesticide applicators
to use to handle this volume of concentrated wastewater. The
technology used by chemical manufacturers for handling large
volumes of low concentrations of wastewater is too expensive to be
feasible for agricultural applicators. Techniques such as
evaporation ponds and gravel disposal pits have been proposed but
these methods require suitable weather conditions for evaporation
and degradation of the pesticides. The overall objective of the
research work that began in 1978 at Purdue was to develop a system
to assist chemical applicators in managing pesticide contaminated

0097–6156/84/0259–0153$06.00/0
© 1984 American Chemical Society

wastewater so that the water could be reused and the   contaminants
could be disposed of in an economical manner.
    The specific objectives were:

1. Investigate alternative methods of removing pesticides from
   contaminated wastewater and design an integrated treatment
   system that would be technically and economically feasible.
2. Demonstrate the system to commercial pesticide applicators and
   provide them with guidelines for installing the system.

The accomplishment of these objectives involved two different
research grants: Grant No. R 805 466010, "Collection and Treatment
of Wastewater Generated by Pesticide Applicators", from the Oil
and Hazardous Spills Branch, U.S. Environmental Protection Agency;
and "Removal of Five R-PAR and Near R-Par Herbicides from
Wastewater", from North Central Regional Pesticide Impact
Assessment Program.
    In the EPA sponsored project (Whittaker, et al. 1982) the
extent of the problem was investigated and alternative means of
removing pesticides from contaminated wastewater were evaluated.
    First the characteristics of typical wash water was measured.
Several aircraft were washed and Table I presents the results of
the analysis of this wastewater.

Table I. Volume and Characteristics of Wastewater Generated by
          Aerial Pesticide Applicators

		Characteristics			
		COD		TSS	SVS
Source	Voume	Total	Soluble		
	liters	mg/1	mg/	mg/1	mg/1
pesticide formulation left in aircraft hopper	5-20	60,000	---	---	---
rinse water used to clean spray boom	40-100	13,000	9,600	11,600	8,900
wash water to clean aircraft hopper	20-40	88,500	5,000	18,000	14,000
wash water to clean aircraft surface	75-200	1,200	500	600	350
Total wastewater	150-360	1,200	900	1,100	950

Next several methods of filtering the pesticide contaminated
wastewater were evaluated. Particle size filters were ineffective
since many of the particles are microcolloidal in size.   Likewise
coalescer type filters were generally inadequate. After studying
these filtration techniques, flocculation procedures were

assessed.   Table   II   presents data from one of the typical tests conducted to determine the result of alum (aluminum sulfate) on sedimentation.

Table II.   Use of Aluminum Sulfate to Remove
Metribuzin by Sedimentation

alum dosage (mg/1)	Initial conc. of Metribuzin (mg/1)	Final conc. of Metribuzin (mg/1)
200	100	81
500	100	90
200	500	330
500	500	315
200	750	585
500	750	585
200	1250	1050
500	1250	1125
200	2000	996
500	2000	982
200	3000	920
500	3000	1000

Alum dosages between 200 and 500 mg/1 were not significantly different.   Table II also illustrates that sedimentation is effective in lowering the concentration of the pesticides to the solubility limit.   Metribuzin is soluble in water to a concentration of about 1200 mg/1 and the sedimentation step reduced the concentration from 2000 and 3000 mg/1 to about 1000 mg/1.   At lower concentrations sedimentation was not effective. Other coagulants and flocculant aids such as hydroxide and ferric chloride, were tested but alum with an anionic polymer (Watcon 1255) was the most effective.

Activated carbon adsorption was selected as the means for the final polishing of the pesticide contaminated wastewater. Filtrasorb 300 (Calgon) was used in these tests.   One of the major questions regarding activated carbon adsorption was the effectiveness of this system on mixed groups of materials. Whittaker (1980) conducted an extensive study to determine how much pesticide could be adsorbed by activated carbon using both isotherm and continuous column systems.   Whittaker used bisolute mixtures of pesticides to determine the effectiveness of 25 gm activated carbon columns in removing combinations of pesticides from wastewater.   Table III shows the results of these studies and indicates that activated carbon can be used to remove mixtures of pesticides.   The lowest exhaustion capacity found in this study was 69 mg of metribuzin adsorbed on 1 gm of activated carbon.

Table III. Capacity Data from Column Adsorption Studies on Bi-Solute Pesticide Solutions

Solution	Concentration (mg/L)	Water Solubility (mg/L)	Breakthrough[1] Capacity (mg/g)	V_B^2 (L)	Exhaustion[3] Capacity (mg/g)	V_E^4 (L)
Ametryne-Propham	160.0	185 (20°C)	76.14	12.0	111.2	23.0
	171.5	250	81.1	12.0	119.2	23.0
Propham-Diphenamid	205.7	250	96.5	11.8	142.0	23.4
	197.4	260 (27°C)	93.0	11.8	136.3	23.4
Fluometuron-Diphenamid	83.7	90 (20°C)	53.0	16.0	88.5	38.0
	178.0	260 (27°C)	102.2	14.5	150.0	28.4
Metribuzin-Propham (Soln. A)	194.3	1220 (20°C)	56.4	7.3	69.0	10.0
	122.0	250	57.0	11.8	-[5]	-
Metribuzin-Propham (Soln. B)	1100	1220 (20°C)	117.8	2.7	151.9	4.5
	189.0	250	29.9	4.0	71.4	15.2
Propham-Monocrotophos	206.0	250	159.5	19.5	-[5]	-[5]
	266.7	misc.	174.7	16.5	186.0	21.0
Monocrotopos-Diphenamid	244.0	misc.	77.4	8.0	109.2	14.0
	167.7	260 (20°C)	65.2	10.0	127.7	28.0
Metribuzin-Monocrotophos	513.3	1220 (20°C)	67.0	3.3	119.5	7.2
	493.9	misc.	52.0	2.65	70.6	7.2
Metribuzin-Methomyl	535.9	1220 (20°C)	65.8	3.1	109.8	7.0
	307.2	58,000	55.8	2.75	72.2	4.40
2,4-D-Propham	272.0	300 g/100g H_2O	64.5	6.0	85.0	9.5
	224.9	250	151.6	17.0	215.4	30.5

[1] Total adsorption capacity where effluent conc. (C_e) equals .1 of influent concentration (C_o)
[2] Volume passed where C_e = .1 C_o
[3] Total adsorption capacity where C_e = .75 C_o
[4] Volume passed when C_e = .75 C_o
[5] The exhaustion capacity of activated carbon for prophram in combination with metribuzin and monocrotophos could not be determined.

Ruggieri (1981) found that a mixture of five herbicides could be removed from contaminated wastewater with the combined treatment system of alum flocculation and sedimentation followed by activated carbon adsorption as shown in Figure 1. The wastewater solution contained approximately 180 mg/l trifluralin, 460 mg/l alachlor, 700 mg/l dinoseb, 180 mg/l paraquat, and 90 mg/l 2,4-D. In these tests trifluralin and paraquat were completely removed through the sedimentation treatment. To remove paraquat, benonite clay was added in the initial sedimentation step. Paraquat strongly attaches to the cation exchange sites on the benonite clay. The trifluralin is essentially insoluble in water and is easily settled after the emulsion is broken.

The procedure followed for flocculation is shown below.

1. Prepare 350 L of synthetic wash water.
2. Add bentonite, alum, and anionic polymer to the synthetic wash water. The dosage was 2 L of powdered bentonite clay, 100 ml of 10 N N_aOH, 20 ml of 57% alum solution and 25 ml of anionic polymer.
3. Mix the solution of wastewater, bentonite, alum, caustic, and polymer for 10 minutes and then flocculate at a mixing speed of about 10 rpm for 30 minutes.
4. Settle for 1 h.

The supernatant from flocculation was then passed through activated carbon columns. The exact concentration of dinoseb, alachlor and 2,4-D used in the study is shown in Table IV. Two carbon column exhaustion studies were conducted. In both cases 300 L of wastewater were pumped through the columns. The carbon adsorbed slightly over 300 mg of pesticide per gram of carbon in both tests. 2,4-D and Dinoseb broke through simultaneously after 1500 L of supernatant with a concentration of about 600 mg/L Dinoseb and 90 mg/L 2,4-D had passed through the carbon. The columns were not totally exhausted until 3000 L had passed through.

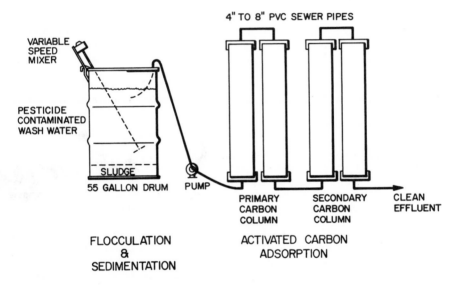

Figure 1. Pesticide-contaminated wastewater treatment system for commercial applicators.

Table IV. Removal of Three Herbicides by the Complete System

Volume	Dinoseb Concentration				Alachlor Concentration				2,4-D Concentration				
1	Inf[1] mg/l	Prim[2] mg/l	Effl[3] mg/l	Adsorbed[4] gm	Inf[1] mg/l	Prim[2] mg/l	Effl[3] mg/l	Adsorbed gm	Inf[1] mg/l	Prim[2] mg/l	Effl[3] mg/l	Adsorbed gm	Efficiency[5] mg/gm
164	940	768	0	126	610	83	0	14	120	84	0	14	25
437	545	331	0	216	340	83	0	36	70	74	0	34	46
710	545	584	0	376	340	56	0	52	70	52	0	48	76
1010	700	653	0	572	440	64	0	71	90	68	0	69	114
1310	700	722	0	788	440	59	0	88	90	56	1	85	154
1501	700	515	8	885	440	88	0	105	90	79	4	100	175
1801	700	722	21	1096	440	83	0	130	90	76	8	122	215
2047	700	561	87	1220	440	54	0	143	90	50	20	133	239
2347	700	480	75	1349	440	59	0	161	90	58	23	147	264
2620	700	389	102	1431	440	54	0	176	90	48	45	154	281
2920	700	561	262	1521	440	57	0	193	90	49	43	155	298
3193	700	384	285	1548	440	59	0	209	90	54	75	158	306
3411	700	377	308	1563	440	55	0	221	90	77		158	310
245	710	798	0	196	440	76	0	19	90	60	0	15	37
554	710	546	0	364	440	66	0	39	90	48	0	29	69
854	730	616	0	549	440	76	0	69	110	56	0	46	105
1154	710	745	0	772	440	68	0	82	90[6]	52	1	62	146
1454	710	630	1	961	440	82	0	107	300[6]	167	1	112	188
1754	710	653	10	1154	440	122	0	143	90[6]	108	3	143	230
2054	710	572	22	1319	440	110	0	176	90[6]	91	9	168	266
2354	710	618	105	1473	1270	143	0	219	110[6]	220	39	223	306
2654	850	639	231	1595	510	126	0	257	155[6]	237	101	263	338
2954	715	641	561	1619	440	108	0	296	155[6]	343	243	294	352
3245	710	789	722	1639	440	116	0	324	100[6]	283	293	291	360

[1] Concentration of pesticide in the original wastewater solution.
[2] Concentration of pesticide after chemical flocculation and sedimentation (primary treatment).
[3] Concentration of pesticides in the effluent after passing through the activated carbon columns.
[4] Amount of pesticide adsorbed into the activated carbon.
[5] Efficiency of activated carbon in terms of total amount of pesticide adsorbed on the carbon.
[6] 2,4-D formulation was changed from Formula 40 (Dow) to Weedone 638 (Am Chem).

Implementation of Treatment System

Following the development of the 2-stage treatment system, demonstrations were performed to show the adequacy of the system at two different pesticide applicators bases within the state of Indiana. Over two spraying seasons the system was used at the Monon Airport to handle the wastewater from ADI, Inc. Wastewater generated during cleaning and loading of aircraft was collected from a concrete pad that was modified so that the water draining off the concrete could be collected in a sump and pumped into a 1000 gallon storage tank. Once a month the collected wastewater was treated with the system. All the pesticides in the wastewater were removed by the sedimentation and activted carbon adsorption process even though a wide variety of chemicals were handled by that applicator. Further studies were conducted at the Capouch Helicopter operation near Rensselaer. Over the past two seasons this system has handled all of the pesticide contaminated wastewater that was produced at that site.

At each of these sites vegetation has developed in areas that had been assumed to be sterile because of contamination by pesticides during previous years.

Generally the procedure that has been developed is effective in reducing the volume of waste that must be handled by a pesticide applicator by a factor of 100. Five thousand gallons of wastewater can be reduced to 50 gallons of sludge and spent activated carbon. Under current regulations these materials would most likely have to be disposed of at a hazardous waste disposal site.

Literature Cited

1. Nye, J. C., Whittaker, K. F. 1980. Collection and Treatment of Rinsewater from Pesticide Application Equipment, Paper No. 80-2108, American Society of Agricultural Engineers, St. Joseph, Michigan.
2. Ruggieri, T. J., 1981, Determination of the ability of a flocculation/sedimentation/activated carbon treatment plant to remove herbicides from application equipment wash water, and examination of the feasibility of bioassays for determination of activated carbon exhaustion. MSAE Thesis, Purdue Univesity, West Lafayette, Indiana.
3. Whittaker, K. F., 1980, Adsorption of selected pesticides by activated carbon using isotherm and continuous flow column system, Ph.D. Thesis, Purdue Univesity, West Lafayette, Indiana.
4. Whittaker, K. F., Nye, J. C., Wukasch, K. F., Squires, R. G., York, A. C., and Kazimier, H. A. 1982. Collection and Treatment of Wastewater Generated by Pesticide Applicators. PB 82-255 365, Oil and Hazardous Materials Spills Branch, MERL-Cincinnati, USEPA, Edison, NY 08837.

RECEIVED April 24, 1984

Long-Term Degradation Studies
Massive Quantities of Phenoxy Herbicides in Test Grids, Field Plots, and Herbicide Storage Sites

ALVIN L. YOUNG[1]

Agent Orange Projects Office, Veterans Administration, Washington, DC 20420

Three long-term studies have been conducted on the fate of 2,4-dichlorophenoxyacetic acid (2,4-D) and 2,4,5-trichlorophenoxyacetic acid (2,4,5-T) when applied in high concentrations to field sites in selected geographical locations. The first study, initiated in April 1970, was of a 208-ha herbicide equipment-testing area (Test Area C-52, Eglin Air Force Base, Florida) that received more than 73,000 kg 2,4,5-T and 76,000 kg 2,4-D during the years 1962-1970. The second study, initiated in 1972, was on the biological degradation of the herbicides when soil incorporated at rates as high as 4,480 kg/ha in plots established in three climatically different areas of the United States; Northwest Florida, Western Kansas and Northwestern Utah. The third study, initiated in 1977, was on the fate of the two herbicides in the soils of two 5-ha sites (Gulfport, Mississippi; and Johnston Island, Pacific Ocean) used for the long-term storage of more than 8.4 million L of surplus phenoxy herbicide. The environmental fate of 2,4-D and 2,4,5-T is compared between the individual studies.

From January 1962 to April 1970, a program of aerial application of herbicides was conducted in Southeast Asia by the United States Air Force (USAF). At the conclusion of this program, considerable amounts of herbicide were left unused.

One of the herbicides used extensively in this project was a herbicide designated as "Agent Orange" which was formulated as a

---

[1]Current address: Office of Science and Technology Policy, Executive Office of the President, Washington, DC 20506

50:50 mixture of the n-butyl esters of 2,4-dichlorophenoxyacetic acid (2,4-D) and 2,4,5-trichlorophenoxyacetic acid (2,4,5-T). In 1970, approximately 8.4 million L of this material were placed in storage by the Air Force. An analysis of the herbicide stocks revealed that it contained the highly toxic contaminant 2,3,7,8-tetrachlorodibenzo-p-dioxin (TCDD). The concentration of the TCDD ranged from <0.02 to 47 ppm TCDD in the 492 random samples taken from the 40,310 208-L drums: the weighted average concentration of TCDD for the inventory was determined to be approximately 2 ppm [1].

Because of the TCDD concentration, the herbicide could not merely be declared surplus and disposed of on the agricultural markets. Hence, the Air Force initiated an extensive research program to find suitable disposal methods that would be both ecologically safe and economically feasible. Although a major method extensively investigated was soil incorporation and bio-degradation, the final disposal method was at-sea incineration, a project conducted in 1977. However, in the course of investigating the feasibility of soil biodegradation, experimental plots were established and sites were studied where the herbicide had been extensively sprayed in the course of developing the spray equipment for Vietnam. When the herbicide was removed from the two storage sites at the time of its destruction, a study of the contamination of those sites was initiated. This paper focuses on the three areas of study that provided data on the environmental fate of 2,4-D and 2,4,5-T in situations where the soil was massively contaminated.

Herbicide Spray Equipment Test Grids

The Eglin Reservation in Northwest Florida has served various military uses, one of them having been the development and testing of aerial dissemination equipment in support of military defoliation operations in Southeast Asia. It was necessary for this equipment to be tested under controlled situations that would simulate actual use conditions as near as possible. For this purpose an elaborate testing installation, designed to measure deposition parameters, was established on the Eglin Reservation with the place of direct aerial application restricted to an area of approximately 3 km^2 within Test Area C-52A in the southeastern part of the reservation. Massive quantities of herbicides, used in the testing of aerial defoliation spray equipment from 1962 through 1970, were released and fell within the instrumented test area. The uniqueness of the area prompted the United States Air Force to set aside the area in 1970 for research investigations. Numerous ecological surveys have been conducted since 1970. As a result, the ecosystem of this unique site has been well studied and documented [2,3].

Although the total area for testing aerial dissemination equipment was approximately 3 km^2, the area actually consisted of four separate testing grids. The primary area was located in the southern portion of the testing area and consisted of a 37 ha instrumented grid. This was the first sampling grid and was in operation in June 1962. It consisted of four intersecting straight lines (flight paths) arranged in a circular pattern, each path being at a 45° angle from those adjacent to it. Although this grid was used from 1962 to 1964, this grid (called Grid I) received 39,550 kg of 2,4-D and 39,550 kg 2,4,5-T as the Herbicide Purple formulation (50 percent n-butyl 2,4-D, 30 percent n-butyl 2,4,5-T and 20 percent iso-butyl 2,4,5-T). Two other testing grids were sprayed with Herbicide Orange. Grid II was an area of 37 ha and located immediately north of Grid I. Grid II received 15,890 kg 2,4-D and 15,890 kg 2,4,5-T from 1964 through 1966. Grid IV was the largest and final grid established on Test Area C-52A. It was approximately 97 ha and received 20,000 kg 2,4-D and 17,570 kg 2,4,5-T from 1968 through 1970. Grid III was an experimental circular grid that received 1,300 kg 2,4-D from 1966 through 1970. Thus, for the four spray equipment calibration grids, a total of approximately 73,000 kg 2,4,5-T and 77,000 kg 2,4-D were aerially disseminated during the period 1962-1970. These data are summarized in Table I.

Table I. Approximate Amount of 2,4,5-T and 2,4-D Applied to Test Area C-52A, Eglin AFB Reservation, Florida, 1962-1970

Test Grid	Grid Area (ha)	2,4,5-T[a] (kg)	2,4-D[a] (kg)
I	37	39,550 (1962-1964)[b]	39,550 (1962-1964)
II	37	15,890 (1964-1966)	15,890 (1964-1966)
III	37	–	1,300 (1966-1970)
IV	97	17,570 (1968-1970)	20,000 (1968-1970)
Total	208	73,010	76,740

[a]Amount of 2,4,5-T and 2,4-D calculated on weight of active ingredient in the military Herbicides Orange and Purple.
[b]Years when the specific grid received the herbicide.

Residue Studies. Despite excellent records as to the number of
missions and quantity of herbicide per mission, there was no way
to determine the exact quantity of herbicide deposited at any
point on the instrumented grids. The first residue studies of
Test Area C-52A involved analyses of soils for phenoxy herbicides
by both chemical and bioassay techniques. These studies,
published by Young (2) in 1974, showed that residues of the
phenoxy herbicides rapidly disappeared. However, problems were
encountered in these residue studies because of the heterogeneity
of the test grids. Not only were there small geologic differences
(soil types, contours, organic matter and pH), and differences
in vegetation density and locations of water, but most important
the herbicides had been sprayed on specific test arrays (i.e.,
along dictated flight paths) over a span of years. An obvious
disparity also existed between bioassay data and chemical
analyses because the latter analysis for 2,4-D and 2,4,5-T alone
could not account for all the biologically active phytotoxic
components. The last application of Agent Orange was applied
in December 1969 at a rate of 28 L/ha. Chemical analyses of
soil cores from the treated areas showed that levels of total
2,4-D and 2,4,5-T in the top 15 cm of soil averaged 2.82 ppm
in April 1970 and less than 8.7 ppb in December 1970.

In October 1973, soil samples collected from Grids I and
II were analyzed and found to contain significant levels of
TCDD. Highest TCDD residues (740-1,500 parts per trillion, ppt)
were found on Grid I, the area sprayed with Herbicide Purple
in 1962-1964. Subsequent soil samples confirmed TCDD
contamination throughout three of the four test grids. The
persistence of TCDD in the soils of Test Area C-52A has recently
been described by Young, 1983 (4).

Vegetative Studies. To demonstrate the rapid dissappearance
of phenoxy herbicides from the environment of the test grids,
a vegetative succession study was conducted of the dicotyledonous
species. Nine months (June 1971) after the last defoliant-
equipment test mission, a detailed survey of the vegetation
was initiated. The 3.0 km^2 area was divided into a grid of
169 sections (each 122 by 122 m), and within each section the
percentage vegetative coverage was visually ranked as Class 0,
0-5%; I, 5-20%; II, 20-40%; III, 40-60%; IV, 60-80%; and V
80-100%. Three sections within each class were selected at
random and surveyed for dicotyledonous plants. An unsprayed
area located 0.3 km northwest of the test area was also surveyed.
In June 1973, each of these areas was again surveyed, but in
addition in 15 sections, nine randomly selected areas, each
0.093 m, were analyzed for species composition and ground
cover density.

Vegetative coverage maps prepared in 1971 and 1973 (Figures
1 and 2 respectively) confirmed that rapid re-vegetation occurred
immediately after herbicide applications ceased. Table II

shows the percent coverage that each vegetative class occupied
in June 1971 and in June 1973.

Table II. Percent of Vegetative Cover Occupied by Vegetative
Class for the 3 km^2 Test Area

Vegetative Class	June 1971	June 1973
0 (0-5%)	4	0
I (5-20%)	14	4
II (20-40%)	29	12
III (40-60%)	25	18
IV (60-80%)	21	42
V (80-100%)	4	23

From June to September 1971, 74 dicotyledonous species were
collected on the 3 km^2 Test Area, and 33 additional species were
found during the June 1973 survey. The most important
dicotyledonous plants found invading the test area were rough
buttonweed, Diodia teres Walt; poverty weed, Hypericum
gentianoides L.; and common polypremum, Polypremum procumbens L.
The studies of soil residues and vegetative succession
of Test Area C-52A confirmed that massive quantities of
phenoxy herbicides rapidly disappeared following the termination
of an aerial spray equipment testing program.

Soil Incorporation/Biodegradation Plots

One potential method proposed for the disposal of Herbicide
Orange was subsurface injection or soil incorporation of the
herbicide at massive concentrations. The premise for such
studies was that high concentrations of the herbicides and TCDD
would be degraded to innocuous products by the combined action
of soil microorganisms and soil hydrolysis. In order to field
test this concept, biodegradation plots were established in
three climatically different areas of the United States;
Northwest Florida (Eglin Air Force Base), Western Kansas (Garden
City) and Northwestern Utah (Air Force Logistics Command Test
Range Complex). A comparison of the soils of the three sites
is given in Table III. The Utah site had a mean annual rainfall
of 15 cm, while the Kansas and Florida sites had 40 and 150 cm,
respectively. Table IV describes the experimental protocol for
the three sites to include when the plots were established, the
method of herbicide incorporation, the experimental design and
the initial calculated herbicide concentration, ppm, at the

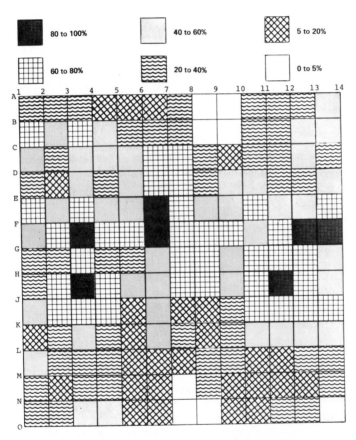

Figure 1. The May 1971 Vegetation Density Map of the 169 sections (each 122 m x 122 m) that constituted the 2.5 km^2 area that received more than 69,300 kg 2,4-D and 2,4,5-T between June 1964 and December 1969, Test Area C-52A, Eglin AFB, Florida.

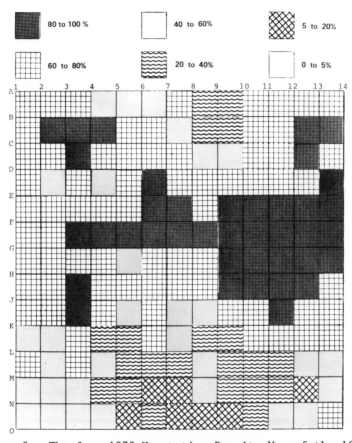

Figure 2. The June 1973 Vegetation Density Map of the 169 sections (each 122 m x 122 m) that constituted the 2.5 km² area that received more than 69,300 kg 2,4-D and 2,4,5-T between June 1964 and December 1969, Test Area C-52A, Eglin AFB, Florida.

time the plots were established. Further details on
experimental protocol can be obtained from Young, et al
(5), 1974, and Young et al (6), 1976.

Table III. Comparison of the Characteristics of the Top 15 cm
           Layer From Each of the Soil Biodegradation Sites

Location	pH	Organic Matter (%)	Sand (%)	Silt (%)	Clay (%)	Soil Description
Eglin AFB, FL[a]	5.6	0.5	91.6	4.0	4.4	Sandy loam
Garden City,KS[b]	7.0	1.7	37	42	21	Silt loam
AFLC Test Range Complex, UT[c]	7.8	1.4	27	53	20	Clay loam

[a]Plots located on Test Area C-52A, Eglin AFB Reservation,
Florida.
[b]Plots located on the Kansas Agricultural Experiment Station,
Garden City, Kansas.
[c]Plots located 120 km west of Salt Lake City, Utah.

Table V compares the degradation of total 2,4-D and 2,4,5-T
(n-butyl esters and acids) over six years of observations in the
Kansas and Florida locations. Although the rates of application
were similar, the method of application, preplant incorporation
versus subsurface injection, resulted in significant differences
in the initial concentrations of herbicides in the plots. The
acid of 2,4,5-T comprised most of the total residue after the
first two years. Although some residues were recovered,
especially in later years, at depths below 15 cm, the majority
(90 percent) of residue was confined to the top 15 cm of soil
profile. The addition of soil amendments such as lime, organic
matter and fertilizer did not appreciably increase the overall
rate of disappearance of the herbicide. The addition of
activated coconut charcoal, however, significantly decreased
the rate of disappearance of herbicide. Six years after the
charcoal plots were established, residues (primarily 2,4,5-T
acid) were still present.
     Microbial studies were conducted on the biodegradation plots
in Florida. Soil samples were taken from all plots in June
and August 1974 (2 years) and in April 1975 (3 years). Although
bacterial and fungal levels were similar for control plots or
plots receiving either herbicide or herbicide plus the soil
amendments lime, fertilizer, and organic matter, the levels
were significantly higher in the plots receiving the activated
charcoal. Microorganisms tended to be concentrated in the level
which contained the charcoal (0-15 cm), but greatly reduced in

Table IV. Descriptions of Three Soil Biodegradation Studies Involving Use of Herbicide Orange

Location	Date Established	Method of Incorporation	Treatment	Calculated Initial Herbicide Concentration (ppm)[c]
Eglin AFB, Florida	4/2/72	Simulated Subsurface Injection (30cm band width)	4,480 kg Herbicide/ha[a]	5,000
			4,480 kg Herbicide/ha, plus soil amendments[b]	5,000
			4,480 kg Herbicide/ha plus soil amendments and activated charcoal	5,000
Garden City, Kansas	5/10/72	Preplant Incorporate (Rototiller)	2,240 kg Herbicide/ha	1,000
			4,480 kg Herbicide/ha	2,000
AFLC Test Range Complex, Utah	10/2/72	Simulated Subsurface Injection (8 cm band width)	1,120 kg Herbicide/ha	5,000
			2,240 kg Herbicide/ha	10,000
			4,480 kg Herbicide/ha	20,000

[a] Rate of herbicide calculated as active ingredient. Herbicide injected at 10-15 cm level. Herbicide or preplant incorporated in the 0-15 cm level. All plots duplicated.

[b] The amendments included 4.5 kg lime, 13.5 kg organic matter, and 1.4 kg fertilizer (12:4:8 for N,P,K, respectively) uniformly mixed within the top 0-30 cm of soil in the plot.

[c] Contained in the top 0-15 cm layer.

Table V. Concentration (ppm) of Total 2,4-D and 2,4,5-T
(Herbicide Orange) Over a Six-Year Period in Field Plots in
Kansas and Florida

| | | Florida[b] | | |
Time After Application (years)	Kansas[a]	Herbicide	Herbicide + Amendments[c]	Herbicide + Amendments + Charcoal[d]
Day 5	1,950	4,900	5700	3,075
0.25	1,070	4,280	5420	2,770
0.5	490	---[e]	---	---
1	210	1,870	2015	---
1.5	40	---	---	---
2.0	<10	508	---	2,660
2.5	---	440	184	---
3.0	---	---	---	---
4.0	---	52	8	---
5.0	---	30	3	120
6.0	---	12	---	360

[a]Garden City, Kansas. Plots established 10 May 1972, 4,480 kg/ha
preplant Incorporated. Data are means of replicate plots,
0-15 cm soil increment.
[b]Eglin AFB, Florida. Plots established 2 April 1972, 4,480 kg/ha
simulated subsurface injection. Data are means of replicate
plots, 0-15 cm soil increment.
[c]The amendments included 4.5 kg lime, 13.5 kg organic matter,
and 1.4 kg fertilizer (12:4:8 for N,P,K, repectively) uniformly
mixed within the top 0-30 cm of soil in the plot.
[d]A 1 cm layer of activated coconut charcoal was applied to the
trench prior to application of the herbicide.
[e]Not analyzed.

number at depths immediately below the charcoal. This effect
of increasing the number of microorganisms may have been due
to adsorption of growth promoting substances (e.g., nutrients
and water) on the surface of the charcoal particles. Although
the number of organisms were greater in these plots, the level
of herbicide residue was also greatest. Apparently, the binding
of the herbicide by the charcoal prevented it from being degraded
by the microorganisms.
    Table VI shows the concentration of herbicide in two of the
three sets of field plots established in Utah in 1972. It was
only after the plots were established and the first soil samples
analyzed that it became apparent that the herbicide formulation

placed in these plots was different than that used in Florida
or Kansas. Indeed, an analysis of the formulation confirmed
the presence of roughly a 50:50 mixture of the n-butyl and
isooctyl esters of both 2,4-D and 2,4,5-T. Note from Table VI
that the n-butyl ester of either 2,4-D or 2,4,5-T disappeared
more rapidly than the isooctyl ester. The hydrolysis of the
isooctyl ester to the acid, probably microbially mediated,
accounts for the presence of the acid.

Table VI. Concentrations (ppm) of the Acid and n-Butyl and
Isooctyl Esters of 2,4-D and 2,4,5-T Placed Subsurface in Utah
Plots

Rate/Date	2,4-D			2,4,5-T		
	n-Butyl	Acid	Isooctyl	n-Butyl	Acid	Isooctyl
1,120 kg/ha						
Initial (1972)	1280[a]	<10[b]	560	770	<10	1230
1975	<10	440	<10	<10	930	40
1978	<10	250	<10	<10	900	20
4,480 kg/ha						
Initial (1972)	5900	<10	2640	3590	<10	5790
1975	10	1970	470	72	1740	3000
1978	<10	1060	95	<10	2900	1080

[a]Data are means of replicated plots.
[b]Detection limit was generally 10 ppm.

Microbial studies have also been conducted on the biodegradation plots in Utah and have been published by Stark et al,
1975 (7). Samples were taken three times throughout the year
(summer, winter, and spring, 1973-1974), and microbial species
present (bacteria, actinomycetes and fungi) were determined.
Bacterial counts were higher for soils with greater moisture
content, but the herbicide, in any concentration, had no
significant effect on the microflora.

As with the studies on the herbicide spray equipment testing
grids at Eglin AFB, Florida, the studies of the biodegradation
plots confirmed the presence and persistence of TCDD. Analysis of
soil samples collected from the Utah plots in 1978 indicated
that 85 percent of the amount of TCDD originally extracted in
1972 could be recovered, suggesting that TCDD applied subsurface
was minimally disappearing.

Studies of Herbicide Storage Sites

During the summer of 1977 the USAF disposed of 8.4 million L
of Herbicide Orange by high temperature incineration at sea.
This operation, Project PACER HO, was accomplished under the
very stringent criteria set forth in an United States
Environmental Protection Agency (EPA) ocean dumping permit.
Among the numerous conditions of the EPA-approved disposal
operation was the requirement for the USAF to conduct extensive
environmental and occupational monitoring of the land-transfer/
loading operations, shipboard incineration operations and
subsequent storage site reclamation and environmental monitoring.
Details of the proposed site monitoring programs were prepared
and approved prior to the disposal of the herbicide. The plan
recommended that soil samples from the storage areas at both
the Naval Construction Battalion Center (NCBC), Gulfport,
Mississippi and Johnston Island, Pacific Ocean, be collected
and analyzed for Herbicide Orange after the completion of
transfer operations. These analyses were to aid in the
establishment of a schedule for future monitoring.
     In July 1977, following the completion of Project PACER
HO dedruming and subsequent site clean-up operations at NCBC
and Johnston Island, Air force scientists initiated an extensive
site monitoring program. The objectives of this program were:
     1. To determine the magnitude of Herbicide Orange
        contamination on the storage areas.
     2. To determine the soil persistence of the two phenoxy
        herbicides contained in Herbicide Orange and the
        dioxin contaminant.
     3. To monitor for any movement of residues from the sites
        into adjacent water, sediments and biological
        organisms.
     In July 1977, a preliminary sampling study was initiated.
This consisted of assessing the heterogenity of the soils on the
sites and the heterogenity of the herbicide concentrations. The
studies conducted on the biodegradation plots showed that
movement of the herbicide components and the TCDD was low; thus
surface sampling, e.g., the top 8 cm of soil, constituted the
primary sampling depth. Twelve sites were selected for sampling
at each location; six were in areas of obvious spills and six in
areas that showed no spill. Not only were the spills discernible
by sight but also by smell. Winston and Ritty (8) had previously
found that the olfactory senses can detect a butyl ester
formulation of 2,4,5-T at levels of 0.4 ppb. The results of
this first sampling after Project PACER HO (1977) are shown
in Table VII. Significant concentrations of herbicides, phenols
and TCDD were detected in soils from spill sites. Variation
in concentrations and in the portion of acids to esters
suggested that the spills were from different time periods.

Accordingly, a more extensive protocol was proposed for future sampling.

Table VII. Concentration (ppm) of Total Herbicides, Total
Phenols, and TCDD in 12 Soil Samples Collected July 1977 from
the Herbicide Orange Storage Areas, Johnston Island and
Naval Construction Battalion Center, Gulfport, Mississippi

Location	Number of Sites	Total Herbicides[a] (ppm)	Total Phenols[b] (ppm)	TCDD (ppm)
Spill Sites				
Johnston Island	8	58,000+42,000	135+120	0.073+0.07
NCBC, Gulfport	6	78,000+42,000	152+ 90	0.24 +0.27
No Spill Sites				
Johnston Island	4	26+15	3+2	NA[c]
NCBC, Gulfport	6	14.2+12.4	<1	NA

[a]Total herbicides refers to concentration of acid and all esters
detected of 2,4-D and 2,4,5-T herbicides. Samples consisted of
top 8 cm of soil.
[b]Total phenols refers to concentration of dichlorophenol and
trichlorophenol.
[c]NA=Not Analyzed.

1978 Protocol. The sites within the two storage areas for
monitoring of residue were determined by whether a spill had
occurred or not occurred at that specific location. The basis for
determining a spill was whether a herbicide stain was discernible
(heavy, light, absent) and whether a herbicide odor was detectable
(strong, mild, absent). Thus, within the storage area numerous
locations were found that had a heavy stain and strong odor
(labeled H/H, presumably representing a recent spill); a light
stain and mild odor (labeled L/L, presumably representing an
older spill); and no stain and no odor (labeled 0/0, presumably
representing an uncontaminated area). Fourteen replications of
each treatment were then randomly selected to represent the
storage area (thus a total of 42 permanently marked sampling
locations at both NCBC and Johnston Island). Twelve of these
locations had been tentatively located and marked in July 1977
with the remaining 30 located and marked in January 1978 with
sampling being conducted on these dates, as well as in November
1978. In collecting the soil samples, a 8 cm square was marked
15 cm away from the site marker pin. At each sampling time soil
was taken from a different "point of the compass" with reference
to the marker pin to insure a fresh and undisturbed profile. At

the designated site a 8X8X8 cm cube of soil was removed with a
ceramic spatula which was rinsed with acetone between uses to
prevent carry-over of residue and microorganisms. Wherever
possible, sediment samples were collected from the drainage
areas in a similar manner.

Results. A summary of the analytical results for the 42 sites
sampled in January and November 1978 for the storage area at
Gulfport, Mississippi is shown in Table VIII. A statistically
significant decrease in the levels of total herbicides and
total phenols was found to occur between the two dates. There
was also a downward trend in TCDD levels, but it was not
statistically different (P.05). This trend in decreasing
levels of TCDD (as well as in herbicides and phenols) is even
more pronounced when the July 1977 data for spill sites (Table
VII) are compared to the 1978 data. Unfortunately, because of
differences in site delineation between 1977 and 1978, data
for spill vs no spill between the 2 years cannot be "paired"
and statistically analyzed. Similar levels of herbicides,
phenols, and TCDD have been found in selected soils of the
Herbicide Orange Storage Area on Johnston Island. Table IX
compares the trends in these compounds over four sampling dates
(August 1977, January and October 1978, and August 1979) from
four sites heavily contaminated with phenoxy herbicide (new
spill sites in 1977). Although herbicide levels significantly
decreased over the periods of sampling, trends for disappearance
of TCDD were not as well defined. The data for these four sites
illustrate the inherent weakness of the sampling protocol. When
a spill occurred on a site, the concentration of chemicals
varied significantly within the spill perimeter. Although the
marker pin for permanently locating the site was placed as near
the center of the spill as possible, that did not necessarily
define the zone of greatest soil contamination. Soil samples
collected over time were collected at different "points-of-the-
compass" around the marker pin. Nevertheless, data for samples
collected at the same site and between other spill sites are
generally of similar magnitude.
    Studies on the penetration of the herbicides and on the
microbial content of the samples were conducted at both the
Naval Construction Battalion Center and Johnston Island. The
results of these studies have been described by Young, et. al.
in 1979 and 1983 (9,10). The data indicated that although
penetration of herbicide and TCDD had occurred throughout the
soil profiles sampled (8 cm increments down to 32 cm), the
bulk of the chemicals remained near the surface. Data from the
microbial analyses of soil samples collected from the Herbicide
Storage Areas confirmed that proliferation of certain microflora
occurred under high levels of herbicide residue.

Table VIII. Mean Concentrations (ppm) of Total Herbicides, Phenols and TCDD in Soils Collected in January and November 1978 from Selected Sites on the Herbicide Orange Storage Area, Naval Construction Battalion Center, Gulfport, Mississippi

Location	Number of Sites Sampled[a]	Total Herbicides (ppm)[b]	Total Phenols (ppm)[c]	TCDD (ppm)
"No" Spills (0/0)[d]				
January 1978	14	32* [e]	3.5*	ND(4)[f]
November 1978	14	3†	0.4†	NA[g]
"Old" Spills (L/L)				
January 1978	14	1,202*	86*	0.0364(3)
November 1978	14	492†	23†	0.0438(3)
"New" Spills (H/H)				
January 1978	14	51,285*	437*	0.2064(10)*
November 1978	14	30,005†	253†	0.1444(11)*

[a]Each soil sample consisted of a cube of soil (8X8X8 cm) removed adjacent to a designated marker.
[b]Total herbicides refers to the concentration of acid and all esters of both 2,4-D and 2,4,5-T.
[c]Total phenols refers to total concentration of both dichlorophenol and trichlorophenol.
[d]The coding 0/0, L/L and H/H are described in the text.
[e]Means within columns within subtitles followed by the same symbols are not significantly different at the 0.05 probability level.
For the statistical analysis, the Wilcoxon Paired-Sample Test was used. A test for a one-tailed hypothesis with paired samples was used in the procedure for nonparametric data since it could not be assumed that the levels of residue detected were from a normal distribution and it was expected that the residues would decrease with time.
[f]ND=Not Detected; the number of samples analyzed is in parentheses. The detection limit was generally 0.0002 ppm (200 ppt).
[g]NA=Not Analyzed.

Table IX. Concentration (ppm) of Total Herbicides, Total Phenols
and TCDD in Soil Samples from Four Selected Spill
Sites for Four Dates from the Herbicide Orange Storage
Area, Johnston Island

Sample Date and Sample Site	Total Herbicides[a] (ppm)	Total Phenols[b] (ppm)	TCDD (ppm)
25 August 1977			
5[c]	38,000	93	0.0330
9	52,270	205	0.0417
10	135,250	460	0.1960
12	76,080	172	0.1780
	75,400	233	0.1122
8 January 1978			
5	38,980	123	0.0340
9	70,090	181	0.0220
10	141,300	477	0.2300
12	57,000	110	0.0800
	76,840	223	0.0915
18 October 1978			
5	31,440	34	0.0191
9	60,530	111	0.0286
10	159,700	456	0.2350
12	42,840	47	0.1110
	73,630	162	0.0984
8 August 1979			
5	3,560	ND	0.0410
9	44,230	149	0.0530
10	48,660	136	0.1300
12	18,430	54	0.0810
	28,720	113(3)[d]	0.0763

[a]Total herbicides refers to concentrations of acid and all esters
detected of 2,4-D and 2,4,5-T.
[b]Total phenols refers to concentrations of dichlorophenol and
trichlorophenol.
[c]The sample consisted of a cube (8X8X8 cm) of soil removed from
near the center of an area designated as a spill.
[d]Refers to number of samples included in obtaining the means.

Discussion and Conclusions

The amount of phenoxy herbicides applied or spilled on a kg/ha
basis in the above studies can only be described as "massive."
Although Grid I on the Eglin AFB spray equipment testing grids
received the herbicides primarily during 1962 and 1963, the
total amount aerially applied was 2,140 kg/ha. Because the
herbicides in this situation were applied from an aircraft, the
time between repetitive applications and the environmental
factors greatly influenced the amount that was incorporated
into the soil profile. Thus, residues were continually disappearing and accumulation and persistence were minimal. However,
in the biodegradation plots and in the Herbicide Storage Areas,
high concentrations of herbicides were applied in a short time
period and incorporated immediately into the soil profile, and
hence, the long persistence time. Nevertheless, these studies
do show that the soil chemistry and the soil microbial populations
can effectively combine to degrade massive concentrations of the
phenoxy herbicides and that recovery of the sites occur as
documented by the re-establishment of the vegetative community.
    The major conclusions from long-term degradation studies of
massive quantities of 2,4-D and 2,4,5-T in test grids, field
plots and herbicide storage areas are:
1. The method of application has significant impact on the
    amount applied per unit area and hence on residue persistence:
            spills > soil incorporation > aerial application.
2. The herbicide 2,4,5-T is more persistent in the soil than
    2,4-D.
3. The formulation of the herbicide has significant impact
    on its persistence:
            isooctyl ester > n-butyl ester > acid
4. The addition of coconut charcoal increases persistence of
    phenoxy herbicide residues, especially residues of 2,4,5-T.
5. The appearance of dichlorophenol and trichlorophenol in soils
    treated with 2,4-D and 2,4,5-T suggests that they are
    degradation products of the herbicides.
6. The massive concentration of herbicides found in these
    studies do not sterilize the soils. Indeed, the data suggest
    that microbial populations respond both quantitatively and
    qualitatively to the presence of high concentrations of
    herbicides and may play a major role in their degradation.
7. The contaminant 2,3,7,8-TCDD has a long persistence time
    in soils (years) and may be a major consideration in the
    use of soil biodegradation as a disposal option for
    "unwanted" phenoxy herbicides or TCDD-contaminated chemical
    wastes.

Acknowledgments

I am indebted to Drs. Eugene Arnold and Mason Hughes for the analyses of the phenoxy herbicides, phenols and the TCDD contaminant.

Literature Cited

1. Young, A. L., J. A. Calcagni, C. E. Thalken and J. W. Tremblay, 1978. "The Toxicology, Environmental Fate and Human Risk of Herbicide Orange and Its Associated Dioxin." Air Force Technical Report OEHL-TR-78-92. 247p. Document available from NTIS, 5285 Port Royal Road, Springfield, VA 22161.
2. Young, A. L., 1974. "Ecological Studies on a Herbicide-Equipment Test Area (TA C-52A), Eglin AFB Reservation, Florida." Air Force Technical Report AFATL-TR-74-12. 141p. Document AD-780 517, available from NTIS, 5285 Port Royal Road, Springfield, VA 22161.
3. Young, A. L., C. E. Thalken and W. E. Ward, 1975. "Studies of the Ecological Impact of Repetitive Aerial Applications of Herbicides on the Ecosystem of Test Area C-52A, Eglin, AFB, Florida." Air Force Technical Report AFATL-TR-75-142. 127p. Document AD-A032 773, available from NTIS, 5285 Port Royal Road, Springfield, VA 22161.
4. Young, A. L., 1983. "Long-Term Studies on the Persistence and Movement of TCDD in a Natural Ecosystem." Environ. Sci. Res. 26:173-190.
5. Young, A. L., E. L. Arnold, and A. M. Wachinski, 1974. "Field Studies on the Soil-Persistence and Movement of 2,4-D 2,4,5-T, and TCDD." Appendix G. Disposition of Orange Herbicide by Incineration. Final Environmental Statement, November 1974. Department of the Air Force, Washington, D.C.
6. Young, A. L., C. E. Thalken, E. L. Arnold, J. M. Cupello, and L. G. Cockerham. 1976. "Fate of 2,3,7,8-Tetrachlorodibenzo-p-dioxin (TCDD) in the Environment: Summary and Decontamination Recommendations." Air Force Technical Report USAFA-TR-76-18. 44p. Document AD-A033 491, available from NTIS, 5285 Port Royal Road, Springfield, VA 22161.
7. Stark, H. E., J. K. McBride, and G. F. Orr, 1975. "Soil Incorporation/Biodegradation of Herbicide Orange. Vol I. Microbial and Baseline Ecological Study of the U.S. Air Force Logistics Command Test Range, Hill AFB, Utah." Document No. DPG-FR-C615F, US Army Dugway Proving Ground, Dugway, Utah 84022, February 1975.

8.   Winston, A. W. and R. M. Ritty, 1971.   "What Happens to
     Phenoxy Herbicides When Applied to a Watershed Area."
     Ind. Vegetation Manage.   4(1):12-14.
9.   Young, A. L., C. E. Thalken, and W. J. Cairney, 1979.
     "Herbicide Orange Site Treatment an Environmental Monitoring:
     Summary Report and Recommendations for Naval Construction
     Battalion Center, Gulfport, Mississippi."   Air Force
     Technical Report OEHL-TR-79-169.   36p.   Document AD-A062
     143, available from NTIS, 5285 Port Royal Road, Springfield,
     VA 22161.
10. Young, A. L., W. J. Cairney and C. E. Thalken, 1983.
     "Persistence, Movement and Decontamination Studies of
     TCDD in Storage Sites Massively Contaminated with Phenoxy
     Herbicides."   Chemosphere 12(4/5): 713-726.

RECEIVED May 1, 1984

Incineration of Pesticide Wastes

THOMAS L. FERGUSON and RALPH R. WILKINSON

Midwest Research Institute, Kansas City, MO 64110

Incineration is one of the preferred methods of
disposal for a wide range of wastes, including
many RCRA-designated hazardous wastes that contain
organic pesticides. Currently, a high level of
activity is focusing on the evaluation and application of incineration technology. This paper presents an overview of those areas of hazardous waste
incineration applicable to pesticide wastes. Research efforts, destruction and removal efficiency
determination, and application of incineration
techniques to pesticide-containing wastes are
reviewed.

Incineration has long been recognized as an effective means for
disposing of unwanted organic pesticides and certain wastes from
pesticide manufacture. (Organometallic pesticides present special problems due to the potential for metal emissions, and are
generally not disposed of by incineration.) As early as 1970, in
the interim guidelines for pesticide disposal by the Federal
Working Group on Pesticides (1), incineration was proposed as one
of the most appropriate methods of pesticide disposal.
     In 1974, federally recommended procedures were published
under authority of the 1972 amendments of the Federal Insecticide,
Fungicide, and Rodenticide Act (FIFRA) that addressed pesticide
disposal (2). These recommendations identified an incinerator
operating at 1000°C (1832°F) with 2-s retention time in the
combustion zone as acceptable for destruction of organic pesticides. Other incinerators, such as those for municipal solid
waste capable of effecting complete pesticide destruction, are
also acceptable. During this same time frame, i.e., from the
early 1970s to date, a number of research and demonstration
studies have been conducted involving pesticide incineration.
Most of these concern either the identification of incinerator

0097-6156/84/0259-0181$06.00/0
© 1984 American Chemical Society

operating conditions required to destroy pesticides or verification that a specific incineration facility is in fact operating
with an acceptable level of destruction efficiency.

While much of this work has addressed the destruction of
pesticides as such, most current activity is concerned with the
more general problem of hazardous waste incineration. The following discussion summarizes some of the key studies that have
dealt specifically with the incineration of pesticides and pesticide wastes, as well as hazardous waste incineration in general.

Pesticide Incineration

Several of the key research projects conducted to assess pesticide incineration are shown in Table I.

Perhaps the single biggest demonstration to date of effective pesticide disposal was the at-sea incineration of Agent
Orange (3). During August 1977, the M/T Vulcanus, operating under a U.S. Environmental Protection Agency (EPA) permit issued
to the U.S. Air Force, incinerated 8.7 million liters (2.3 million gallons) of Agent Orange in the mid-Pacific Ocean. Destruction efficiency was estimated to be at least 99.99%, and no
detectable 2,3,7,8-tetrachlorodibenzo-p-dioxin (TCDD) was found
in the stack gas.

General Electric Company, under an EPA permit, incinerated
nearly 6,000 L (1,500 gal.) of 20% liquid DDT formulations in a
liquid injection incinerator near Pittsfield, Massachusetts, in
September 1974 (4). The facility utilized a vortex combustor of
the type normally used for disposal of oils and solvents. Operating temperatures ranged from 870 to 980°C with retention times
of 3 to 4 s and 120 to 160% excess air. Overall destruction
efficiency exceeded 99.99%. Concentrations of DDT, DDE, and DDD
in the stack gas and scrubber water were below analytical detection limits.

In a similar study, Versar, Inc., in 1974-1975, demonstrated for EPA that DDT and 2,4,5-T formulations were destroyed in a municipal sewage sludge incinerator in Palo Alto,
California (5). The pesticides were added to sludge (which contained 20% by weight of solids) to form a mixture that was 2 to
5% by weight in pesticides. Destruction efficiencies ranged
from 99.95 to 99.99% for an average hearth temperature from 600
to 690°C and an afterburner temperature from 650 to 660°C.

A two-step laboratory thermal-decomposition analytical system involving vaporization and thermal destruction was developed
in 1975 by the University of Dayton Research Institute for EPA
(6). Vaporization of pure pesticide occurred at 200 to 300°C and
was followed by decomposition in a quartz tube at temperatures
exceeding 900°C. The destruction efficiencies for DDT, Kepone,
and mirex exceeded 99.99% at 2-s residence time and greater than
900°C.

Table I.   Selected Pesticide Incineration Studies

Pesticides	Investigators	System	DE or DRE (%)
Agent Orange	Ocean Combustion Services	Liquid Injection	> 99.99
DDT	General Electric Co.	Liquid Injection	> 99.99
DDT and 2,4,5-T	Versar, Inc.	Sewage Sludge	99.95- 99.99
DDT, Kepone, and Mirex	Univ. Dayton Research Inst.	Two-Stage Vaporization/ Destruction	> 99.99
Aldrin, Atrazine, Captan, DDT, Malathion, Mirex, Picloram, Toxaphene, and Zineb	Midwest Research Institute	Multiple Chamber	> 99.99
Chlordane, DDT, Dieldrin, 2,4-D, Lindane, and 2,4,5-T	TRW Systems, Inc.	Multiple Chamber	> 99.99
Chlordane and Hexachlorobenzene	Rockwell Intl.	Molten-Salt Bath	> 99.99
PCP-Treated Wooden Boxes	Los Alamos National Lab.	Multiple Chamber	> 99.99

Note: DE = destruction efficiency;
        DRE = destruction and removal efficiency.

Subsequently, modified thermal decomposition systems were developed by the University of Dayton to investigate the destruction of complex hazardous waste mixtures (7). These systems utilized various combinations of gas chromatographs/mass spectrometers/computers which were incorporated into the vaporization and decomposition process.

Using these systems, the thermal decomposition of DDE, DDT, diazinon, endrin, hexachlorobenzene, Kepone, mirex, and pentachloronitrobenzene was studied, and in several instances stable intermediate products of incomplete destruction were observed. Kepone at 400 to 500°C yielded hexachlorocyclopentadiene, hexachlorobenzene, and an unidentified chlorinated hydrocarbon, all

of which were stable at these temperatures. Similarly, pentachloronitrobenzene at 500 to 550°C yielded hexachlorobenzene
which was thermally decomposed at temperatures above 650°C.

In 1974 Midwest Research Institute operated a pilot-scale
multiple chamber incinerator to evaluate for EPA the operational
variables for pesticide incineration (8). The system included a
pilot-scale incinerator, a three-stage scrubber system, and a
scrubber water treatment system. Nine pesticides (aldrin,
atrazine, captan, DDT, malathion, mirex,' picloram, toxaphene,
and zineb) in 15 liquid and solid formulations were studied.
Destruction efficiencies generally exceeded 99.99% over a range
of temperatures and retention times (~ 950 to 1100°C, ~ 1.2 to
~ 6 s, and 80 to 160% excess air). This study also documented
the generation of measurable quantities of cyanide in the incinerator off-gas during the incineration of organonitrogen
pesticides.

TRW Systems, Inc., conducted a laboratory-scale incineration study for the U.S. Army from 1973 to 1975 (9). Eleven individual pesticide formulations and three mixed pesticide formulations containing six different active ingredients (chlordane,
2,4-D, DDT, dieldrin, lindane, and 2,4,5-T) were incinerated in
a liquid injection incinerator. The experimental apparatus consisted of a fuel atomizer, combustion chamber, afterburner,
quench chamber, and scrubber unit. Destruction efficiencies
exceeded 99.99% for a minimum 0.4-s residence time at temperatures above 1000°C with 45 to 60% excess air.

Molten salt is a technique that has been considered for the
destruction of pesticides and other hazardous wastes for several
years. In a recent study by Rockwell International for EPA
(10), the destruction of solid hexachlorobenzene (HCB) and liquid chlordane exceeded 99.99% in a molten sodium carbonate bath
at 900 to 1000°C with a residence time of 0.75 s. For the
pilot-scale tests, the concentration of HCB and chlordane in the
spent melt was < 1 ppm. The HCl concentration in the off-gas
was < 100 ppm.

A prototype system (100 kg waste per hour) which utilizes a
nonrefractory metal vessel is under construction and was sched-
uled to begin operating in late 1983 (10). The unit is designed
for transport to field sites for temporary waste destruction such
as in cleanup of a hazardous waste site or continuous waste
destruction at a waste generator's plant site.

In 1981 the Los Alamos National Laboratory investigated for
EPA the thermal destruction of wooden boxes treated with penta-
chlorophenol (PCP). The incineration system consisted of a
dual-chamber, controlled-air incinerator, a spray quench column,
a venturi scrubber, and a packed-column acid gas absorber (11).
Destruction efficiencies for PCP exceeded 99.99% for combustion
chamber temperatures above 980°C, 20% excess air, and a retention time greater than 2.5 s. For these conditions, TCDD and

tetrachlorodibenzofuran were undetectable in the off-gases at
analytical detection limits of 1 ppb and 5 ppb, respectively.

RCRA Regulations

Interest in the general topic of hazardous waste incineration
has been stimulated by the Resource Conservation and Recovery
Act (RCRA) of 1976. Under this authority, EPA has enacted regulations affecting facilities which generate, store, treat, and
dispose of hazardous waste. Incineration has received a great
deal of attention in this regulatory process because it is a
technology that is already available and provides permanent disposal for many types of organic wastes.

A waste stream may be defined as hazardous under RCRA if it
meets certain criteria for ignitability, corrosivity, reactivity,
or toxicity, or if the waste stream is specifically identified by
EPA as a hazardous waste. An individual waste stream is subject
to being classified as hazardous (listed) if it contains any one
of approximately 375 chemicals identified by EPA as hazardous
constituents. These designated chemicals are frequently referred
to as Appendix VIII compounds because of where they are listed in
the published regulation.

Pesticides are very much a part of the definition of hazardous wastes (Table II). In fact, the toxicity characteristic
of hazardous waste as defined by RCRA (referred to as extraction
procedure or EP toxicity) is based on threshold concentrations of
eight metals and six pesticides in an extract of the waste (Table
II-A). Sixteen of the specific hazardous waste streams listed by

Table II.   Pesticides in RCRA Hazardous Waste

A. EP (Extraction Procedure) Toxicity

2,4-D	Lindane	Silvex
Endrin	Methoxychlor	Toxaphene

B. Specific Manufacturing Wastes

Cacodylic Acid	2,4-D	Phorate
Chlordane	Disulfoton	2,4,5-T
Creosote	MSMA	Toxaphene

C. Example Appendix VIII Hazardous Constituents

Aldrin	DDT	Endosulfan	Parathion
Amitrole	Dieldrin	Endrin	PCNB
Chlordane	Dimethoate	Heptachlor	Toxaphene
Creosote	Disulfoton	Methoxychlor	Warfarin

Source: 40 CFR Part 261.

EPA result from the manufacture of nine specified pesticides (Table II-B). Of the approximately 375 Appendix VIII chemicals, about one-fifth are pesticide active ingredients; some examples are shown in Table II-C. Thus, compliance with RCRA incinerator regulations is directly related to the incineration of pesticides.

Recent efforts have been concerned with the acceptable operation of hazardous waste incinerators. Currently, the performance standards for incinerators burning hazardous waste address three areas:

- First, the incinerator must achieve a destruction and removal efficiency (DRE) of at least 99.99% for each of the designated Appendix VIII chemicals present in the waste feed. In other words, $\leq$ 0.01% of the respective compound in the waste feed can be emitted in the incinerator stack gases. The specific Appendix VIII chemicals evaluated are selected by EPA from those found in the hazardous waste at reasonable concentrations and are termed principal organic hazardous constituents (POHCs).
- Second, HCl emissions must not exceed either 1.8 kg/h (4 lb/h) or 1% of the HCl in the stack gas prior to entering any pollution control equipment, whichever is greater.
- Third, the particulate matter emitted must not exceed 180 mg/DSCM (0.08 gr/DSCF), when corrected to 7% O_2.

Most of the ongoing activity relating to hazardous waste incineration (and therefore pesticide incineration) is focused on identifying facilities capable of meeting these requirements and on verifying their performance.

Hazardous Waste Incineration

Incineration of hazardous wastes in an acceptable manner requires a rather sophisticated facility in terms of the actual incinerator, its associated control systems, and the pollution control devices. Existing industrial production facilities as well as dedicated hazardous waste incinerators have been investigated for potential application.

Industrial Processes. Table III shows several high temperature production processes (designated industrial furnaces) that have been considered for the incineration of hazardous waste (12). Criteria considered in evaluating the use of such facilities include the compatibility of the process with hazardous waste disposal, the number and availability of appropriate facilities, and the proximity of the facilities to sites generating hazardous waste.

Process compatibility must include consideration of operating conditions such as temperature and retention time, any actual or perceived impact on product quality, and the potential

Table III. High Temperature Processes
Selected for Further Evaluation

Industry	Process or Furnace
Brick	Tunnel Kiln
Carbon Black	Oil Furnace Process
Primary Copper	Reverberatory Furnace
Primary Lead	Blast Furnace
Iron and Steel	Blast Furnace
	Open Hearth Furnace
Lime	Long Rotary Kiln
	Short Rotary Kiln with
	Stone Preheater
Glass	Melting Furnace

for emission of hazardous wastes during materials handling operations (fugitive emissions). Based on these criteria, the EPA
initiated a study to identify potential high temperature processes. Initially, most major processes with an operating temperature greater than 649°C (1200°F) were included.

There are both technical and institutional problems, however, in using most of these processes for waste incineration.
While the potential for obtaining low cost energy does exist,
there seems to be very limited potential for the use of most of
these facilities for incinerating pesticide wastes.

The wet process cement kiln is in a somewhat different category from other high temperature industrial processes because
cement kilns have been used to incinerate several chemical wastes
and their DREs have been rather extensively evaluated (13-15).
Cement kilns are well matched to hazardous waste incineration
considering their size (up to 25 ft in diameter and over 500 ft
long) and operating temperatures of 1370° to 1450°C (2500° to
2650°F) in the kiln burning zone. Another important factor is
their ability to adsorb chlorine emissions in the cement, thus
aiding in meeting the HCl emission control requirement (16).

Tests have been conducted on wet and dry cement processes
in the United States, Canada, Sweden, and Puerto Rico for wastes
containing a wide range of chlorinated chemicals, including PCBs
(13-16). Generally, the DREs have been found to be in the range
of 99.99%, with no adverse impact on product quality or plant
operation if chlorine addition is restricted to less than 1%
of the net fuel/waste feed. DREs of less than 99.99% have been
observed, however, where poor control of combustion air exists
and waste is inadequately atomized, even when an acceptable
cement product is being produced.

Burning wastes as fuel can have a significant economic impact on wet process cement production facilities, where fuel

costs can represent as much as 65% of the cost of cement clinker production, and may help extend the economic life of this older process (16).

Hazardous Waste Incinerators. Major efforts have been applied to the evaluation of existing industrial and commercial hazardous waste incinerators (17). The number of these incinerators, including those under construction, is estimated to be 392 (see Table IV). These facilities include several types of incinerators, although over half are the liquid injection type. If one considers the geographical distribution of these facilities on the basis of EPA regions, the greatest number are in Regions 4 and 6. Texas and Louisiana are the two states with the greatest number of facilities. (Information on operational hazardous waste incinerators in specific locations can be obtained from EPA regional offices.)

Table IV. Projected Hazardous Waste
Incinerator Population

Type	Number
Liquid Injection	213
Hearth	75
Rotary Kiln	17
Fluidized Bed	5
Other	42
Under Construction	40
Total	392

Currently, there are two principal areas of activity involved in the evaluation of hazardous waste incineration facilities. Both are primarily concerned with DRE determination and with HCl and particulate control. The first area concerns EPA's efforts to finalize the operating permits for hazardous waste incineration facilities. The second phase (Part Bs) of the operating permits for many hazardous waste incinerators are being called in by EPA. The completed permit application must include documentation that the incinerator is capable of meeting the prescribed operating conditions, i.e., a $\geq$ 99.99% DRE, and HCl and particulate control. Thus, many of these facilities have already conducted test burns or are planning tests soon. The test results will eventually be made public through the permitting process.

The second area of activity has to do with studies being conducted by EPA to evaluate incinerator performance as part of the Agency's Regulatory Impact Analysis (RIA) of hazardous waste incineration regulations. As part of this analysis, several facilities burning hazardous waste have been tested by EPA. Data from these tests are being analyzed to determine compliance with the operational requirements. The results of this study, however, have not yet been published.

One incinerator that has been evaluated rather extensively and for which test results have been reported is the liquid chemical waste incinerator facility owned by the Metropolitan Sewer District (MSD) of Greater Cincinnati, Ohio (18). The MSD facility uses a rotary kiln and liquid injection cyclone furnace to incinerate a wide variety of liquid industrial chemical wastes. The total design heat release rate is 120 million kJ/h (114 million Btu/h). Tests conducted over a wide temperature range (~ 900°C to ~ 1300°C) for six Appendix VIII chemicals (carbon tetrachloride, chloroform, hexachlorobenzene, hexachlorocyclopentadiene, and hexachloroethane) have shown DREs equal to or very near 99.99%.

In addition to these full-scale tests, EPA has initiated a program to conduct extensive intermediate-scale incinerator studies, i.e., studies that would approximate the actual conditions that exist in full-scale incinerators but that at the same time would be close enough to the laboratory studies previously discussed to allow correlation of the results from both scales of operation (19). The EPA Combustion Research Facility (CRF) has been constructed to conduct this program at the National Center for Toxicological Research (NCTR), Jefferson, Arkansas.

Summary

Over the past 15 years, significant effort has been expended in evaluating pesticide incineration. Some of the earliest work on industrial waste incineration was performed on pesticides, and this work became part of the basis for the current regulations for the incineration of hazardous wastes.

While earlier studies addressed the incineration of pesticides and pesticide wastes as such, most current efforts are focused on the general area of hazardous waste, as defined by the Resource Conservation and Recovery Act of 1976. This ongoing work is directly related to pesticide disposal, however, as pesticide waste is included in the category of RCRA hazardous waste. In fact, the presence of pesticides is a major consideration in a waste being designated as hazardous.

Significant activity is occurring in assessing hazardous waste incineration in both the private and public sectors. Much of the information gained from this effort will be directly applicable to our knowledge of the incineration of pesticides and pesticide-containing wastes.

Literature Cited

1. Working Group on Pesticides. "Summary of Interim Guidelines
    for Disposal of Surplus or Waste Pesticides and Pesticide
    Containers," Report WGP-DS-1, 1970.
2. U.S. Environmental Protection Agency. "Pesticides and Pes-
    ticide Containers: Regulations for Acceptance and Recom-
    mended Procedures for Disposal and Storage," Fed. Regist.
    1974, 39(85), 15236.
3. Ackerman, D. G.; Fisher, H. J.; Johnson, R. J.; Maddalone,
    R. F.; Matthews, B. J.; Moon, E. L.; Scheyer, K. H.; Shih,
    C. C.; Tobias, R. F. "At-Sea Incineration of Herbicide
    Orange Onboard the M/T Vulcanus," EPA-600/2-78-086, 1978.
4. Leighton, I. W.; Feldman, J. B. "Demonstration Test Burn of
    DDT in G.E. Liquid Injection Incinerator," U.S. Environmen-
    tal Protection Agency, 1975.
5. Whitmore, F. C. "A Study of Pesticide Disposal in a Sewage
    Sludge Incinerator," EPA/530/SW-116c, 1975.
6. Duvall, D. W.; Rubey, W. A. "Laboratory Evaluation of High
    Temperature Destruction of Kepone and Related Pesticides,"
    EPA-600/2-76-299, 1976.
7. Graham, J. L.; Rubey, W. A.; Dellinger, B. "Determination
    of Thermal Decomposition Properties of Toxic Organic Sub-
    stances," Presented at the National Meeting of American In-
    stitute of Chemical Engineers, Cleveland, Ohio, August
    1982.
8. Ferguson, T. L.; Bergman, F. J.; Cooper, G. R.; Li, R. T.;
    Honea, F. I. "Determination of Incinerator Operating Condi-
    tions Necessary for Safe Disposal of Pesticides," EPA-600/
    2-75-041, 1975.
9. Shih, C. C.; Tobias, R. F.; Clausen, J. F.; Johnson, R. J.
    "Thermal Degradation of Military Standard Pesticide Formu-
    lations," TRW Systems, Inc., 1975.
10. Johanson, J. G.; Yosim, S. J.; Kellogg, L. G.; Sudar, S.
    "Elimination of Hazardous Wastes by the Molten Salt De-
    struction Process," Proc. 8th Annu. Res. Symp.: Incinera-
    tion and Treatment of Hazardous Waste, EPA-600/9-83-003,
    1983.
11. Stretz, L. A.; Vavruska, J. S. "Controlled Air Incineration
    of PCP-Treated Wood," Los Alamos National Laboratory/U.S.
    Environmental Protection Agency IERL/CI Incineration Re-
    search Branch, U.S. EPA Interagency Agreement AD-89-F-
    1-539-0, undated.
12. Hall, F. D.; Kemner, W. F.; Staley, L. J. "Evaluation of
    Feasibility of Incinerating Hazardous Wastes in High-
    Temperature Industrial Processes," Proc. 8th Annu. Res.
    Symp.: Incineration and Treatment of Hazardous Waste, EPA-
    600/9-83-003, 1983.

13. Higgins, G. M.; Helmstetter, A. J. "Evaluation of Hazardous Waste Incineration in a Dry Process Cement Kiln," Proc. 8th Annu. Res. Symp.: Incineration and Treatment of Hazardous Waste, EPA-600/9-83-003, 1983.
14. Black, M. W.; Swanson, J. R. Pollut. Eng. 1983, 15(6), 50.
15. Chadbourne, J. F. "Burning Hazardous Waste in Cement Kilns," Presented at the 76th Annual Meeting of the Air Pollution Control Association, Atlanta, Georgia, June 1983.
16. Lauber, J. D. J. Air Pollut. Control Assoc. 1982, 32(7), 771.
17. Keitz, E.; Boberschmidt, L. "A Profile of Existing Hazardous Waste Incineration Facilities," The MITRE Corp., November 1982.
18. Ananth, K. P.; Gorman, P.; Hansen, E.; Oberacker, D. A. "Trial Burn Verification Program for Hazardous Waste Incineration," Proc. 8th Annu. Res. Symp.: Incineration and Treatment of Hazardous Waste, EPA-600/9-83-003, 1983.
19. Carnes, R. A.; Whitmore, F. C. "Siting and Design Consideration for the Environmental Protection Agency Combustion Research Facility," Proc. 8th Annu. Res. Symp.: Incineration and Treatment of Hazardous Waste, EPA-600/9-83-003, 1983.

RECEIVED March 26, 1984

TECHNOLOGY DEVELOPMENT

A Large Scale UV-Ozonation Degradation Unit
Field Trials on Soil Pesticide Waste Disposal

PHILIP C. KEARNEY, QIANG ZENG, and JOHN M. RUTH

Pesticide Degradation Laboratory, Beltsville Agricultural Research Center, U.S. Department of Agriculture, Beltsville, MD 20705

Decomposition of farm-generated pesticide wastewater was demonstrated with a mobile 66-lamp ultraviolet (UV) unit and ozone. Aqueous solutions of 2,4-D (1086 ppm) and atrazine (4480 ppm) were degraded more than 80% in about 2-3 h, while paraquat (1500 ppm) was degraded more slowly. Dwell time, or the time the molecule was actually in the lamp unit, and concentration were two parameters that affected the rate of degradation. Mass spectra of the trimethylsilyl (TMS) derivatives of atrazine subjected to UV-ozonation revealed a number of dehalogenated, dealkylated s-triazines, paraquat yielded the 4-picolinic acid, and 2,4-D gave oxalic acid, glycolic acid and several four-carbon oxidation products. The economics of UV-ozonation as a pretreatment for land disposal compares favorably with incineration and other options open to the small pesticide user.

A frequent problem encountered by the pesticide user is safe waste disposal of liquid chemicals generated during or subsequent to a spray operation. No accurate data are available on the total magnitude of the pesticide wastewater generated annually in the United States, but some information is available from one segment of the agricultural community. Commercial aerial applicators spray about two-thirds of all applications made on agricultural and forest lands or roughly half of all pesticide applications in the U.S. (1). It has been estimated that in a normal year 10,000 aircraft sprayed 180 million acres once and 350-380 million acres for those crops that receive more than one application. It is also estimated that 10-60 gallons of wastewater are generated each day per spray plane, with concentrations ranging from 100-1000 ppm (2,3). Based on Seiber's estimates (4), using 30 gallons and 500 ppm, it may be

calculated that roughly 10,000 gallons of wastewater containing
20 kg pesticide are generated each year per plane. Estimates
beyond this are probably meaningless, because not all spray
planes are engaged in pesticide application. Nevertheless, we
suspect the overall problem of safe wastewater is substantial,
based on the fact that more than 225.1 x 10^6 kg (active
ingredient) of pesticide were used on major field crops in the
U.S. in 1982, of which 191 x 10^6 kg were herbicides, 24.4 x
10^6 kg insecticides, 2.4 x 10^6 kg fungicides, 0.4 x 10^6 kg
miticides, 1.9 x 10^6 kg fumigants, 2.2 x 10^6 kg defoliants,
and 2.7 x 10^6 kg plant growth regulators (5).

Wastewaters are often generated at scattered or remote
sites, making transportation and approved decontamination
difficult and costly. Faulty waste disposal techniques can also
potentially make a substantial contribution to ground water
contamination. A need exists, then, to develop new technology to
solve a major problem faced by the pesticide user that has
potential impact both within and outside of the agricultural
community. Recently we have reported on the use of ultraviolet
(UV) irradiation, in the presence of ozone, as a pretreatment
prior to soil disposal of aqueous pesticide solutions (6-8).
Oxidative pretreatment was found to render a number of
chlorinated compounds more biodegradable when exposed to the
natural soil microflora. These studies were conducted primarily
with a 450 W medium-pressure mercury vapor lamp using
^{14}C-labeled compounds to monitor the breakdown process both
during UV-ozonation and in soils after addition of the oxygenated
products. In addition, a commercial lamp unit containing 66 low-
pressure mercury lamps was used for some experimental runs on
larger volumes of chemicals.

The objective of the research reported here was to determine
whether the 66-lamp unit would detoxify waste pesticide solutions
generated on a farm with high daily pesticide usage.

Methods and Materials

Location. The location of the experiments was at the Farm
Operations Division at USDA's Beltsville Agricultural Research
Center (BARC). A diverse farming system is in operation at
Beltsville, including both experimental plots and large acreages
devoted to forage and grain production for our livestock
research. About 3000 acres are sprayed annually to control
weeds, insects, and diseases. Arrangements were made with the
farm manager to supply us with excess pesticide solutions
remaining in the spray tank at the end of a spraying operation.
These solutions were stored in 10-gallon stainless steel milk
cans and processed when time permitted, but usually within
several days after receipt.

UV-Unit. The large unit is an Ultra-Violet Purifier manufactured
by Pure Water Systems, Inc., 23 Madison Road, P. O. Box 1387,
Fairfield, New Jersey 07006. The unit consists of 66
low-pressure mercury vapor lamps with a maximum energy output at
253.7 nm of 455 W. Each lamp is encased in a long quartz tube
and the tubes are arranged within a stainless steel cylinder,
approximately 40 cm in diameter so that each lamp is located 1.27
cm from each adjacent lamp (Figure 1). The volume of the lamp
chamber is 1 cubic foot (7.48 gallons or 28.32 L). Liquid is
delivered to the lamp unit by a pump with a flow rate of 8 to 40
L/min. A large stainless steel holding tank (ca. 210 L) is
connected to the pump, and the lamp unit is connected to the
holding tank, so that the liquid can be recycled through the lamp
unit.
        The unit was mounted on a 8 x 18 foot trailer, tightly
secured, and moved some 5 miles to the pesticide mixing and
loading area on the BARC research facility (Figure 2).
Inspection of the unit during travel and at the farm site showed
no major problems in making it mobile for transport to other
sites for further research and development activities.

Chemicals processed. Waste pesticide solutions were collected
after spray operations during late May, June, and July of 1983,
and consisted primarily of three compounds: 2,4-D [(2,4-dichloro-
phenoxy)acetic acid], atrazine (2-chloro-4-(ethylamino)-6-
(isopropylamino)-s-triazine) and paraquat (1,1'-dimethyl-4,4'-
bipyridinium dichloride). Our efforts were primarily directed at
these pesticides, which are shown in Table I together with their
formulations and concentrations.

Table I. Pesticides, Formulations and Concentrations Subjected
          to UV-Ozonation in the 66-Lamp Unit

Pesticide	Formulation	Concentration (ppm)
2,4-D	isooctyl ester, 4EC (low volatile ester)- 68.2% ai, 31.8% inerts	1086
Atrazine	Aatrex 4L – 40.8% ai, 2.2% related compounds 57.0%	12000
Paraquat	29.1% ai, 70.9% inerts	1500

        Routinely 38 to 152 L of wastewater were added to the large
holding tank (210 L capacity) and allowed to recycle through the

Figure 1. Lamp chamber containing 66 low-pressure mercury vapor lamps encased in quartz tubes, each situated 1.27 cm from adjacent lamp.

Figure 2. Mobile 66-lamp UV unit showing holding tank, pump and lamp chamber mounted on a trailer.

system for about 30 min with no irradiation. Some water remains in the light chamber from the previous day's cleaning; therefore, this equilibration time was necessary to establish a uniform sample at time 0. Prior to activating the lamps, pure oxygen was fed into the light chamber at a rate of 100 cc/min, and this rate was maintained throughout the processing time. In some studies pure ozone was fed into the light chamber from model GTC-1B ozone generator (Griffin Technics Corp. 66 Route 46, Lodi, New Jersey 07644) at a rate of 32 g per h from O_2 feed gas (manufacturer's estimate). Samples were taken periodically and returned to the Pesticide Degradation Laboratory for chemical analyses.

Residue analysis. All analyses were conducted by gas chromatography (glc) using a flame ionization detector and a 5% QF-1 column on Chromsorb W (DMCS) (2 mm i.d. by 1.8 M), 80/100 mesh; N_2 flow rate at 40 mL/min and column temperatures of 160, 215, and 180°C for atrazine, 2,4-D, and paraquat, respectively.

For atrazine, 10 mL of the commercial formulation (5000 ppm) was extracted 3 times (3x15 mL) with ethylacetate (EtAc). The EtAc was dried with Na_2SO_4, filtered, reduced in volume under N_2, and adjusted to 10 mL for glc analysis. A sample spiked with 0.1 μCi (U ring-^{14}C) atrazine yielded an extraction recovery of 99.6% via scintillation counting. A similar extraction procedure was used for 2,4-D, with a recovery of 76.0%.

For paraquat, 10 mL of the commercial formulation (1500 ppm) were taken to dryness in a rotary evaporator, 300 mg of $NaBH_4$ and 15 mL of 95% ethanol were added, and the solution was heated for 15 min at 60°C, cooled to room temperature, and carefully taken to dryness in a rotary evaporator. Fifteen mL of distilled water were added, and the solution was extracted 3 times (3x15 mL) with hexane. The hexane was dried with Na_2SO_4, filtered, reduced in volume with a gentle stream of N_2 and adjusted to a final volume of 10 ml for glc analysis. A sample spiked with 0.1μCi (methyl -^{14}C) paraquat yielded an extraction recovery of 68.8%.

Laboratory studies. Small-scale laboratory studies were conducted with the 450 W Hanovia lamp unit to investigate the products of UV-ozonation, which were analyzed on a gas chromatograph-mass spectrometer. Purified aqueous samples of atrazine, paraquat, and 2,4-D were irradiated for 30 min in the presence of oxygen, the water was removed under N_2, and trimethylsilyl (TMS) derivatives were prepared using bis(trimethylsilyl)trifluoroacetamide. A portion of the TMS derivative was injected into a 30-M flexible silica capillary column, coated with SE-30, and temperature programmed from 90 to 200°C at 5°C/min. The mass spectrometer is a Finnigan 4021 with an Incos data system, operated in the election impact mode, with an electron energy of 70eV and a source temperature of 250°C.

Results and Discussion

The results will be discussed in two sections. The first section
will deal briefly with our laboratory findings. The second
section will deal with the on-farm disposal results.

Laboratory studies. A chromatogram of the total ion current of
the TMS derivatives of irradiated atrazine is shown in Figure 3.
Peak assignments were based on reference standards and previous
mass spectra of TMS derivatives of 6 suspected microbial metabolites of simazine [2-chloro-4,6-bis (ethylamino)-s-triazine] a
related s-triazine (9). The region between scan 1250 and scan
1600 contained large peaks for derivatives of hydroxyatrazine,
N-ethylammeline, N-isopropylammeline and ammeline. The largest
peak was the di TMS derivative of hydroxyatrazine with a
molecular ion at M/Z 341. The loss of chlorine in the
UV-ozonation process appeared to be complete, since no chlorine
containing compounds were detected. Next in abundance are
derivatives of compounds obtained by removing the ethyl,
isopropyl or both from hydroxy atrazine and replacement by
hydrogen. Based on the mass spectral data, the major products
are shown in Figure 4.
     If the number of active hydrogens in the compound is greater
than the number of attached TMS groups, then isomeric forms of
the derivative are possible. That situation is well illustrated
by two mono TMS derivatives of hydroxyatrazine, represented by
the small glc peak at scan 1288 and the much larger peak at scan
1335. The latter is presumably the TMS-O-derivative and the
other a TMS-N-derivative.
     The tri-TMS peaks for ammelide, N-ethyl ammelide, and
N-isopropyl ammelide, with maxima at scans 1096, 1186, and 1208,
respectively, are very small in comparison with the peaks between
scan 1250 and scan 1600.
     The compounds giving rise to peaks at scan 734 to 744, scans
746 to 747, and scans 815-816 can be explained by assuming that
one amino group on hydroxyatrazine has been removed entirely and
replaced by a hydrogen atom. The first two overlap to give what
looks like a single large glc peak at scan 744. In Figure 3, the
proposed names of those compounds have been enclosed in brackets
to indicate that the elemental composition is probably correct
but that this structure has not yet been verified. It is
possible that the rings have been opened. Similar structures can
be postulated for nearly all of the small peaks between scan 400
and scan 700, but at least one requires a ring alteration. These
peaks, plus the ones between scan 200 and scan 400 and those
between scan 1500 and scan 2000, require further study.
     The TMS derivative of 4-picolinic acid (isonicotinic acid)
was identified in the TMS-treated reaction mixture from the
UV-ozonation of paraquat. This suggests a sequence of reactions,
in which Slade (10), Funderburk et al. (11), and others have made

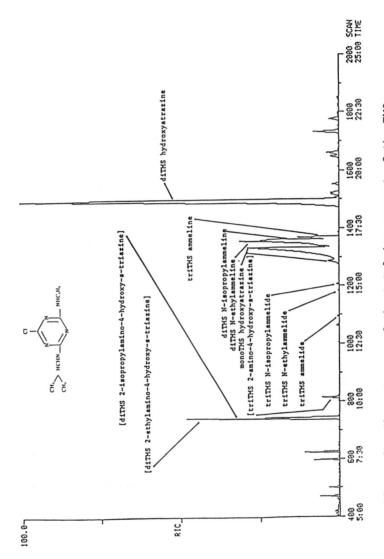

Figure 3.   Chromatogram of the total ion current of the TMS derivatives of atrazine subjected to UV-ozonations in a 450 W medium-pressure mercury vapor lamp.

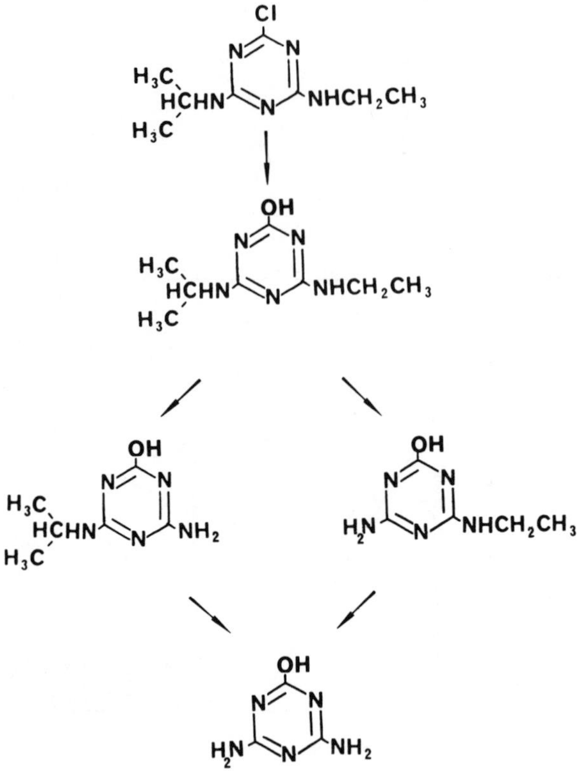

Figure 4.   Products resulting from UV-ozonation of atrazine.

early, major contributions (Figure 5). UV-ozonation of the
4-picolinic acid yielded a number of products, which are
currently under investigation.

With 2,4-D, neither the trimethylsilyl ester nor the free
acid was found among the reaction products. The dominant
chromatographic peaks in the mixture represent the TMS
derivatives of glycolic acid, oxalic acid, and several
four-carbon oxidation products. Some of the latter have been
tentatively identified, and the work needed to confirm choices
between alternative structures is continuing. Some smaller
chromatographic peaks probably represent derivatives of
structures containing five or more carbon atoms. The nature of
the products suggests ring fragmentation of 2,4-D, as opposed to
the products derived from atrazine. Unpublished $^{14}CO_2$
evolution studies with (^{14}C-ring) 2,4-D confirm extensive
ring oxidation of this compound during UV-ozonation.

Most of the products identified in these studies are
considerably more biodegradable than the parent materials and
should be degraded or bound more rapidly than the parent
compounds in biologically active soils.

Field studies. Our research with the large UV reactor on the
Beltsville Research Farm was directed toward finding the optimal
conditions for destroying pesticide wastewater using
UV-ozonation. We did not concentrate on the microbial phase at
this time, other than in a very limited way.

Figure 6 shows the decomposition of 38 and 152 L of 2,4-D at
1086 ppm. The larger volume (152 L) took about 5 times longer to
achieve about a comparable degree of degradation, ie. slightly
over 80%. The difference would appear to be due to dwell time,
or the time the molecule actually spends in front of the light
source as opposed to the time in the 210 L holding tank. Volume,
then, is one of the variables that can be manipulated, and most
of the subsequent research was done in the 38 L or 10-gallon
range. The effect of generated ozone on 2,4-D degradation is
shown in Figure 7. High levels of ozone did not substantially
accelerate the initial oxidation of 2,4-D, but it did increase
the rate of degradation in more dilute solution during the latter
part of the process.

Our early results with atrazine indicated that this molecule
was somewhat more stable than 2,4-D (Figure 8). Preliminary
results suggested that the initial concentration of atrazine
(12000 ppm) may have an effect on the rate of reaction. We
studied the effects of dilution in our small lamp unit to
determine whether lowering the concentration might accelerate the
rate of degradation (Figure 9). As the concentration decreased
by a factor of 10, the time required for degradation also
drastically decreased. The difference between 120 and 12 ppm is
indistinguishable. We attempted to use this finding in the large
unit by adding 114 L of water to the holding tank and then slowly

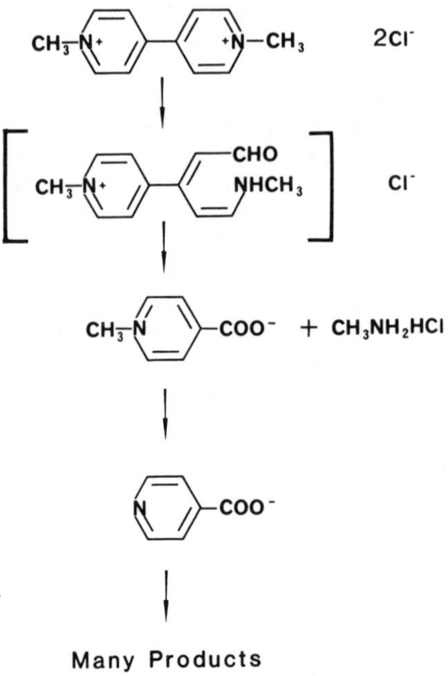

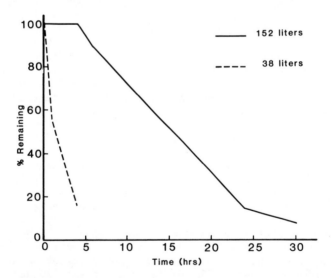

Many Products

Figure 5. Suggested reaction sequence for oxidative decomposition of paraquat based on the identification of picolinic acid by mass spectrometry.

Figure 6. Effect of volume size on the rate of degradation of formulated isooctyl ester of 2,4-D at a concentration of 1086 ppm in the 66-lamp unit.

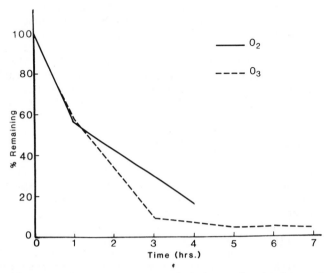

Figure 7. Effect of O_2 and O_3 fed into the 66-lamp chamber during UV-irradiation of formulated isooctyl ester of 2,4-D at a concentration of 1086 ppm.

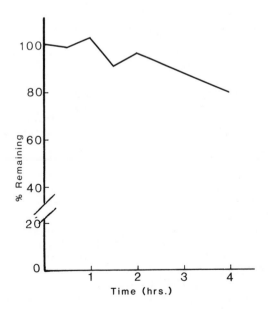

Figure 8. Degradation of atrazine at 12000 ppm via UV-ozonation in the 66-lamp unit.

adding 38 L of more concentrated atrazine to the water in the
reservoir at a rate of 44 ml/min. The initial concentrations
would be extremely dilute and thus more rapidly decomposed. We
could calculate the theoretical concentration at each time period
and measure actual concentration by glc (Figure 10). As in
previous runs, oxygen was fed into the lamp chamber. The results
were disappointing, since the atrazine showed no appreciable
breakdown during the mixing phase when the concentration would
have been reasonably dilute. The large volume (152 L) and the
consequently lower dwell time are probably responsible for the
slow rate of decomposition.

To optimize the conditions, ozone from the Griffin generator
was fed into the lamp chamber in a non gradient run and at a
concentration of 4480 ppm. When we used the same air flow rate
for ozone as for oxygen, ie. 100 cc/min, we noted no striking
difference in the rate of degradation. If, however, we operated
the ozone generator at its maximum output, a very rapid decline
was measured (Figure 11). The initial decomposition was rapid
and was essentially complete after 2 h.

Our research progress with paraquat has not been as
extensive as with 2,4-D and atrazine. Preliminary results
(Figure 12) suggest breakdown was slower than anticipated. At
this point it is meaningless to compare the rates of breakdown of
the three herbicides, since the concentrations and formulations
are quite different.

Two critical parameters that must be controlled for the
process to be effective in a reasonable time period are
concentration and dwell time. These parameters will vary with
each instrument, depending on design and capacity of the various
units.

Two important areas of research are underway to improve
further the hybrid UV-O_3/soil disposal system. A UV monitor
coupled with a computer is under investigation to make the unit
self contained with regard to chemical analyses and to automate
certain functions of the photochemical phase. Second, a
microbial enrichment investigation utilizing UV-O_3 products
as substrates is in progress to select and possibly engineer
degradators to accelerate the microbial phase.

Economics. The manufacturer's retail price for the 66-lamp unit
from Pure Water Systems, Inc. is $35,000. The cost of operation
is based on energy usage, which is 1.5 KWH (manufacturer's
estimate); if the current cost of electricity is 5¢/KWH, then the
daily operation would run $1.80/day or $657 annually. The cost
of oxygen and the ozone generator are not included, nor do we
have any estimate on long-term maintenance.

It is difficult to compare costs to other disposal options.
The Seiber report (4) listed the capital cost for incineration at
$89,550, and yearly operating cost of $106,500, based on 1978
estimates. The capital and operating costs for our UV-ozonation

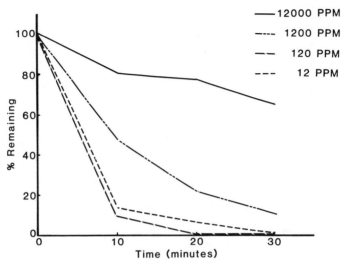

Figure 9. Effect of concentration on the degradation of formulated atrazine at four concentrations via UV-ozonation in a 450 W medium-pressure mercury vapor lamp.

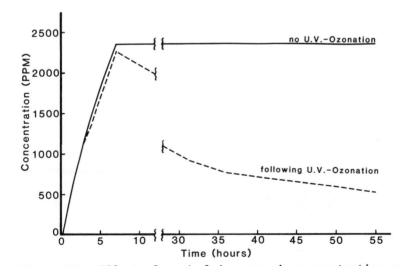

Figure 10. Effect of gradual increase in concentration up to a concentration of 2356 ppm on the degradation of formulated atrazine up to a concentration of 2356 ppm via UV-ozonation in the 66-lamp unit.

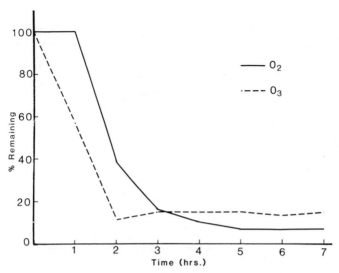

Figure 11. Effect of O_2 and O_3 fed into the 66-lamp unit during UV-irradiation of formulated atrazine at 4480 ppm.

Paraquat U.V.-Ozonation 1500 PPM

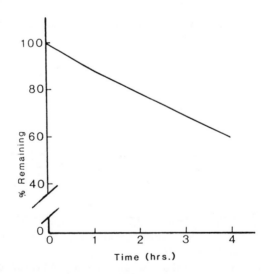

Figure 12. Degradation of paraquat via UV-ozonation at 1500 ppm in the 66-lamp unit.

would appear to be more favorable than incineration, but not as economical as physical treatment and the several land disposal options.

The advantages of the unit would be its mobility; relative ease of operation; low operating cost; production of less toxic, biodegradable products; and assurance of extensive degradation in a relatively short period of time when compared to ground disposal.

Use of a company or product name by the Department does not imply approval or recommendation of the product to the exclusion of others which may also be suitable.

Literature Cited

1. Collins, H., personal communication.
2. "Disposal of Dilute Pesticide Solutions," U.S. Environmental Protection Agency Report SW-176C (NTIS Report PB-297 985) 1979.
3. Whittaker, K. F.; Nye, J. C.; Wukarch, R. F.; Squires, R. G.; York, A. C.; Kazimier, H. A. "Collection and Treatment of Wastewater Generated by Pesticide Applicators," U.S. Environmental Protection Agency.
4. Seiber, J. N. "Disposal of Pesticide Wastewater-Review, Evaluation and Recommendations," U.S. Environmental Protection Agency, OER, ORD, Draft Report 1981.
5. Schaub, J. R. The Economics of Agricultural Pesticide Technology (in press) in J. L. Hilton, ed. Agricultural Chemicals of the Future. Vol. 8. Beltsville Symposia in Agricultural Research. 1983.
6. Kearney, P. C.; Plimmer, J. R.; Li, Z-M. 181st Am. Chem. Soc. Natl. Mtg. Abstr. PEST. 48, Atlanta, Georgia. 1981.
7. Kearney, P. C.; Plimmer, J. R.; Li, Z-M. Proc. 5th Int. Congr. Pestic. Chem. IUPAC 1983. 4,397.
8. Kearney, P. C.; Quiang, Z.; Ruth, J. M. Chemosphere (in press) 1983.
9. Lusby, W. R.; Kearney, P. C. J. Agric. Food Chem. 1978, 26, 635.
10. Slade, P. Nature 1965, 207, 515.
11. Funderburk, H. H.; Negi, N. S.; Lawrence, J. M. Weeds 1966, 14, 240.

RECEIVED April 16, 1984

Reaction of Sodium Perborate with Organophosphorus Esters

GRACE LEE, RICHARD A. KENLEY[1], and JOHN S. WINTERLE

SRI International, Menlo Park, CA 94025

The objective of this work is to establish a reaction mechanism between sodium perborate and several organophosphorus esters. By analogy we can then describe its probable effects upon other phosphorus-based insecticides. We conclude that the reactivity of sodium perborate toward five model compounds is attributable to the nucleophilic reactions of hydroperoxyl anion, HO_2^-, produced by perborate dissociation in water. On this basis we predict that sodium perborate solutions will be effective chemical detoxicants for phosphorus ester insecticides.

Sodium perborate (PB) is a commercially available, storage stable powder that dissolves in water to liberate hydrogen peroxide ($\underline{1}$), and in alkaline solution, HO_2^-. The perhydroxyl anion is extremely reactive towards phosphorus esters (OP) ($\underline{2}$), and has been proposed as a reagent for treating toxic wastes ($\underline{3}$). From the known constants for PB/hydrogen peroxide ($\underline{4}$) and hydrogen peroxide/perhydroxyl ($\underline{5}$) anion equilibria we estimated that saturated PB solutions would rapidly degrade water-solubilized OP insecticides at pH 10 or below. However, the literature provides no evidence for reaction rates or mechanism or reaction of PB with OP compounds; therefore its reactivity requires experimental evaluation.

Our hypothesis is that PB solutions owe their great reactivities with OP chemicals to the highly nucleophilic perhydroxyl anion, HO_2^-, produced by the dissociation of PB to HO_2^- and boric acid. We have tested this hypothesis by calculating bimolecular substitution rate constants, k_{HOO}, for

[1]Current address: Syntex Corporation, Palo Alto, CA 94304

0097-6156/84/0259-0211$06.00/0
© 1984 American Chemical Society

HO_2^- from our reaction kinetic data obtained for PB solutions
containing different OP components. These numbers are then
compared with literature values for k_{HOO}, with H_2O_2 as the source
of HO_2^-.

PB was reacted with the following model compounds: paraoxon
(diethyl p-nitrophenyl phosphate), EPMP (ethyl p-nitrophenyl
methylphosphonate), PDEP (p-nitrophenyl diethylphosphinate, PNPA
(p-nitrophenyl acetate), and PMP (hydrogen p-nitrophenyl
methylphosphonate).

Experimental Materials. Sodium perborate tetrahydrate
($NaBO_3 \cdot 4H_2O$) was obtained from Alfa Products; titration for
available H_2O_2 by $KMnO_4$ gave > 95% of theoretical equivalency.
The following chemicals were obtained from the indicated supplier
and used without further purification: paraoxon and sodium
perchlorate, Aldrich Chemical Company; sodium borate (anhydrous)
and sodium phosphate dibasic (anhydrous), Malinckrodt, Inc.;
disodium ethylene diamminetetraacetate, Fisher Scientific Company;
sodium hydroxide, J. T. Baker, Company; tetraethyl biphosphine
disulfide, Chemical Procurement Laboratory; diethyl methylphosphonate, Specialty Organics, Inc. PMP, from Ash-Stevens, Inc.
PMP was purified by repeated crystallization until exhaustive
alkaline hydrolysis of PMP samples showed quantitative (> 99%)
liberation of p-nitrophenolate (PNP). Water for kinetic
experiments was obtained from a Milli-Q reverse-osmosis/ion
exchange system and was glass-distilled before use.

We synthesized EPMP, by the method of Fukuto and Metcalf (6)
by reacting diethyl methylphosphonate with phosphorus penta-
chloride, followed by reaction with sodium p-nitrophenolate. EPMP
was purified by distillation at reduced pressure. Analysis for
$C_9H_{12}NO_5P$. Calculated: % C, 44.1; % H, 4.93; % N, 5.71. Found: %
C, 43.9; % H, 4.77: % N, 4.59. [Caution! EPMP is a VERY TOXIC
NERVE POISON; the subcutaneous lethal dose (LD50) in mice is
350 μg/kg (R. Howd, R. Kenley, unpublished), and the material
should be handled with extreme care at all times.]

PDEP, was synthesized by reacting tetraethyl biphosphine
disulfide with $SOCl_2$ by the method of Parshall, (7) yielding
$(C_2H_5)_2P(O)Cl$. Subsequent reaction of the phosphinochloridate
with p-nitrophenol following the general method of Douglas and
Williams (8) yielded the desired product, which was purified by
vacuum distillation. Analysis for $C_{10}H_{14}NO_4P$. Calculated: % C,
49.28; % H, 5.74; % N, 5.74. Found: % C, 49.08; % H, 5.83; % N,
5.59.

Safety Precautions. Because the phosphorus esters used in our
studies exhibit varying (and sometimes unknown) degrees of acute
toxicity, we recommend observing strict laboratory safety
precautions when handling these materials.

Apparatus. Spectrophotometric determination of PNP production was performed with a Perkin-Elmer Model 554 uv-visible spectrophotometer equipped with a thermostatted 5 x 5 position cell holder and cell programmer. The cell programmer automatically cycles between cuvette positions and permits the unattended operation of the instrument. Temperature control was maintained at $27.5 \pm 0.2°C$ with a Forma Scientific circulating constant-temperature bath. Reaction temperatures were determined in the cuvettes with a National Bureau of Standards calibrated thermometer. Quartz 1-cm pathlength cuvettes were used throughout. pH readings and adjustments were made using a potentiometric Metrohm Model E526 automatic titrater/pH meter.

Methods. All experiments were conducted in aqueous solution with PB in at least 10-fold molar excess over the other reactant. Under these conditions we expected and observed pseudo-first order kinetic production of PNP from OP esters. Kinetic solutions also contained the following: 0.1×10^{-3} mol dm^{-3} disodium ethylene diamminetetraacetate, 0.1 mol dm^{-3} buffer (sodium borate unless otherwise specified), and $NaClO_4$ added to bring the solution to ionic strength = 0.50 mol dm^{-3}.

Typically, PB was added to the appropriate buffer medium and used immediately to minimize any possible loss of peroxygen content. OP reagents were transferred via microliter syringe to a second buffer solution; as for PB, solutions were used immediately after preparation. The reactions were initiated by transferring an appropriate volume of PB solution (via Pipetman automatic pipettor) to a cuvette and then similarly adding OP ester solution to bring the total reaction volume to 3.00 cm^3.

PNP production was monitored at 402 nm and quantitated using extinction coefficients determined experimentally for each reaction medium. The fraction of reactant conversion to product was given by the ratio $(A_t - A_o)/(A_\infty - A_o)$ where the subscripts t, o, and ∞ refer, respectively, to absorbance values taken at time t, initially, and at long reaction times when PNP liberation clearly stopped.

Rate constants were determined by linear least-squares regression analysis, and error limits are reported as standard deviations (S.D.).

The reaction mechanism between PB and OP compounds is complex (see equations (2) through (7) below) and leads to very cumbersome algebraic expressions; we chose an alternate technique to calculate α, the dissociation fraction of PB, needed in our kinetic analysis below.

The reaction set was numerically modeled using the computer program CHEMK (9) written by G. Z. Whitten and J. P. Meyer and modified by A. Baldwin of SRI to run on a MINC laboratory computer. CHEMK numerically integrates a defined set of chemical rate equations to reproduce chemical concentration as a function of time. Equilibria can be modeled by including forward and reverse reaction steps. Forward and reverse reaction rate

constants were chosen in ratios to give the following equilibria
constants based on literature values (4,5,10) extrapolated to
27.5°C: $K_2 = 1.2 \times 10^{-4}$, $K_3 = 2.5 \times 10^{-12}$, $K_4 = 6.6 \times 10^{-11}$. The
calculations were precise because concentrations obeyed both mass
balance equations and mass action expressions for equations (2)
through (4) (see below). The constant α was calculated from the
expression $\alpha = ([PB]_o - [PB])/[PB]_o$.

Results

Table 1 shows reactant concentrations ($[PB]_o$, $[PDEP]$), fractional
product conversion ($[PNP]_\infty/[PDEP]_o$), observed half-time ($t_{1/2}$),
and pseudo-first-order rate constant (k_{obs}) values for reaction of
PDEP in pH 8 buffer with and without added PB.

Table I. Experimental Data for Reaction of Sodium
Perborate with PDEP at pH 8 (Run 4953-22)

	Cuvette 1	Cuvette 2	Cuvette 3	Cuvette 4	Cuvette 5
$10^3 [PB]_o$ (mol dm^{-3})	0	0.500	1.00	2.50	5.00
$10^6 [PDEP]_o$ (mol dm^{-3})	5.08	4.98	4.88	4.57	4.07
$[PNP]_\infty/[PDEP]_o$	1.08	1.08	1.08	1.09	1.07
$10^{-4} t_{1/2}$ (s)	2.50	1.47	0.852	0.540	0.336
$10^5 k_{obs}$, (s^{-1})[a]	2.77	4.72	8.12	12.9	2.07

[a] k_{obs} calculated according to equation (1).

The k_{obs} values were calculated according to equation (1):

$$-\ln[(A_t - A_o)/(A_\infty - A_o)] = k_{obs} \cdot t \tag{1}$$

The data of Table 1 show, that for $[PB]/[PDEP] > 10$, the reaction
conversions were quantitative, and that PB concentrations of 0.5
to 5 mM significantly accelerated the liberation of PNP from
PDEP. These data are typical of all experiments performed to
date, so details for other reactions are not reported here.
    Figure 1 is a plot of kinetic data according to equation (1)
for reaction of EPMP with various concentrations of PB at pH =
8. Figure 1 is typical for all other experiments performed and

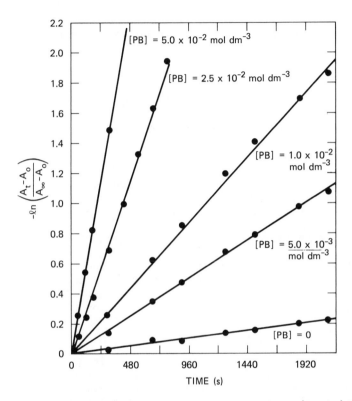

Figure 1. Pseudo-first-order kinetic plot of - $n(A_t-A_0)/(A-A_0)$ versus time for production of p-nitrophenolate from reaction of sodium perborate at various concentrations with EPMP at 27.5 °C, pH = 8.

shows that the data adhere to equation (1) at least to 90% conversion.

Finally, Figure 2 shows k_{obs} plotted as a function of $[PB]_o$ for paraoxon solutions, repeated at three pH's. These data are again representative, showing that k_{obs} is linearly related to $[PB]_o$.

For the simultaneous reaction of added OP ester with base, water, and perhydroxyl, we must consider the following minimum reaction set:

$$PB \; \underset{}{\overset{K_2}{\rightleftharpoons}} \; HO_2^- + B(OH)_3 \tag{2}$$

$$H_2O_2 \; \underset{}{\overset{K_3}{\rightleftharpoons}} \; H^+ + HO_2^- \tag{3}$$

$$B(OH)_3 + H_2O \; \underset{}{\overset{K_4}{\rightleftharpoons}} \; B(OH)_4^- + H^+ \tag{4}$$

$$HO_2^- + OP \; \overset{k_{HOO}}{\longrightarrow} \; PNP + products \tag{5}$$

$$OP \; \overset{k_{sp}}{\longrightarrow} \; PNP + products \tag{6}$$

From this scheme we can obtain equation (7):

$$-d[OP]/dt = [OP](k_{sp} + k_{HOO}[HO_2^-]) \tag{7}$$

in which k_{sp} is the pseudo-first-order rate constant for OP hydrolysis in the absence of PB. With PB in great excess, in buffered solution, k_{sp} and $[HO_2^-]$ are constant. Then pseudo-first-order kinetics result and we get equation (8),

$$-\ln([OP]_t/[OP]_o) = k_{obs} \cdot t \tag{8}$$

if we define $k_{obs} = k_{sp} + k_{HOO}[HO_2^-]$. This is equivalent to equation (1) used in our spectrophotometric assay.

At any pH, $[HO_2^-]$ will depend on the fraction of PB dissociated (α) and the fraction of the total peroxide content existing as HO_2^-. This latter number is given by $K_3/(K_3 + H^+)$. Combining these results gives:

$$[HO_2^-] = (\alpha)[PB]_o \, K_3/(K_3 + [H^+]) \tag{9}$$

From our definition of k_{obs} we finally get by substitution of (9):

$$k_{obs} = k_{sp} + k_{HO_2}(\alpha[PB]_o K_3/(K_3 + [H^+])) \tag{10}$$

Equation (10) predicts the dependence of k_{obs} on $[PB]_o$, illustrated in Figure (1), and requires that the slope be equal to $k_{HOO}(\alpha K_3/(K_3 + [H^+]))$ and that the intercept be k_{sp}.

Since α can be calculated, and K_3 and $[H^+]$ are known, we can determine k_{HOO} from the slope of a linear regression of k_{HOO} as a

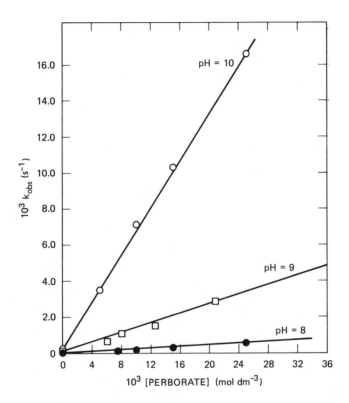

Figure 2. Kinetic plot of pseudo-first-order rate constant for reaction of paraoxon (k_{obs}) versus concentration of added sodium perborate at 27.5 °C at various pH in 0.1 mol dm⁻³ borate buffer.

function of $[PB]_o$. This was done for five compounds at pH 8. The results are displayed in Table 2 along with literature values for k_{HOO}.

Table II. Comparison of Experimental and Literature
Values for k_{HOO}

Compounds	Slope $(mol^{-1} dm^3 s^{-1})^a$	$k_{HOO} (mol^{-1} dm^3 s^{-1})$	Reference
Paraoxon	$(2.2 \pm 0.1) \times 10^{-4}$	1.1×10^0	this work
		8.4×10^{-1}	11
EPMP	$(1.0 \pm 0.1) \times 10^{-2}$	5.1×10^1	this work
PDEP	$(3.2 \pm 0.2) \times 10^{-2}$	1.6×10^2	this work
PMP[b]	$(5.9 \pm 0.1) \times 10^{-4}$	2.0×10^{-3}	this work
		2.6×10^{-3}	12
PNPA	$(5.6 \quad 1.1) \times 10^{-1}$	2.9×10^3	this work
		3.7×10^3	13

[a]From a linear regression of k_{obs} as a function of $[PB]_o$. By our analysis $k_{HOO} = (slope) (K_3 + [H^+])/K_3)(\alpha)^{-1}$. At pH 8.00, $\alpha = 0.78$ and $(K_3 + [H^+])/K_3 = 4.0 \times 10^3$.
[b]In pH 12.00, 0.10 mol dm^{-3} Na_2HPO_4 buffer. $\alpha = 1.0$ and $(K_3 + [H^+])/K_3 = 1.4$.

Discussion

Aqueous PB solutions are well-behaved kinetically with the five OP's studies, as evidenced by excellent pseudo-first-order kinetics and mass balance between OP and PNP at the reaction termination. Our kinetic analysis confirms that reactions (2) through (6) adequately describe PB reactivity, judged by the good agreement of calculated k_{HOO} values with literature values.
    Equation (10), upon which our analysis is based, implicitly assumes that α is constant, independent of $[PB]_o$. This condition must be true as demonstrated by the linear relationship between k_{obs} and $[PB]_o$. Our modeling efforts (not shown) confirm this result. It is a consequence of performing reactions in excess borate buffer, which results in a buffering of $[PB]$.
    In agreement with others, (13) we find HO_2^- to be 50-fold more reactive than ^-OH as a nucleophile toward electrophilic phos-

phorus, and, therefore, sodium perborate is a possible reagent for chemical detoxification of pesticide wastes because it supplies significant concentrations of HO_2^- in the pH range 8 to 12.

Conclusions

Sodium perborate, dispersed in water, enhances the degradation rate of phosphorus esters. It owes its reactivity to hydroperoxyl anion, a powerful nucleophile, which is produced by dissociation of PB in aqueous solution. Because of its stability, commercial availability, and great reactivity we recommend PB as a detoxicant for hazardous OP wastes.

Acknowledgements

This research was supported by the United States Army Armament Research and Development Command through contract DAAK11-82-K-0003.
    Approved for public release; distribution unlimited.

Literature Cited

1. Kirk-Othmer Encyclopedia of Chemical Technology, 2nd ed., Vol. 14, (Wiley Interscience, 1967), pp. 758-760.
2. Edwards, J. O.; Pearson, R. G.; J. Amer. Chem. Soc. 1962, 84, 16.
3. Edwards, J. O.; Proceedings of Symposium on Non Biological Transport and Transformation of Pollutants on land and Water, National Bureau of Standards, 1976.
4. Edwards, J. O.; J. Amer. Chem. Soc. 1953, 75, 6154.
5. Evans, M. G.; Uri, N.; Trans. Faraday Soc. 1949, 45, 224.
6. Fukuto, T. R.; Metcalf, R. L.; J. Amer. Chem. Soc. 1959, 81, 372.
7. Parshall, G. W.; Org. Syn. 1965, 45, 102.
8. Douglas, K. T.; Williams, A.; J. Chem. Soc. Perkin II, 1976, 515.
9. CHEMK, A Computer Modeling Scheme for Chemical Kinetics (undated). G. Z. Whitten and J. P. Meyer. Systems Applications, Inc., 950 Northgate Drive, San Rafael, CA 94903.
10. (a) Determination of pH. Theory and Practice, R. G. Bates (Wiley Interscience, 1973), p 125.
    (b) Quantitative Chemistry. J. Waser (W. A. Benjamin, 1964), p. 400.
11. Epstein, J.; Demek, M. M.; Rosenblatt, D. H.; J. Org. Chem. 1956, 21, 796.
12. Behrman, E. J.; Biallas, M. J.; Brass, H. J.; Edwards, J. O.; Isaks, M.; J. Org. Chem. 1970, 35, 3069.
13. Fina, N. J.; Edwards, J. O.; Int. J. Chem. Kin. 1973, 5, 1.

RECEIVED February 13, 1984

Abiotic Hydrolysis of Sorbed Pesticides

D. L. MACALADY

Department of Chemistry and Geochemistry, Colorado School of Mines, Golden, CO 80401

N. L. WOLFE

Environmental Research Laboratory, U.S. Environmental Protection Agency, Athens, GA 30613

The hydrolysis of pesticides which are sorbed to sterilized natural sediments has been investigated in aqueous systems at acid, neutral and alkaline pH's. The results show that the rate constants of pH independent ("neutral") hydrolyses are the same within experimental uncertainties as the corresponding rate constants for dissolved aqueous phase pesticides. Base-catalyzed rates, on the other hand, are substantially retarded by sorption and acid-catalyzed rates are substantially enhanced. A large body of evidence will be presented which substantiates these conclusions for a variety of pesticide types sorbed to several well-characterized sediments. The significance of our results for the evaluation of the effects of sorption on the degradation of pesticides in waste treatment systems and natural water bodies will also be discussed.

Whether such disposal is intentional or incidental, significant quantities of pesticides and pesticide wastes end up in natural and artificial aquatic systems. Thus, any consideration of the disposal of this broad category of anthropogenic chemicals must include an understanding of the reaction mechanisms and principal pathways for degradation of pesticides in aquatic systems. Of the degradative pathways relevant to such systems, hydrolysis reactions are perhaps the most important type of chemical decomposition process (1-7).

Since many pesticides are compounds of low water solubility, their form in aquatic systems is often dominated not by material in aqueous solution, but rather by material sorbed to suspended or bottom sediments (8-9). Thus, an understanding of the hydrolytic reactions of pesticides which are sorbed to

sediments is crucial for an adequate representation of the
dominant chemical degradative pathways for these compounds in
aquatic systems.

It is the purpose of this article to summarize the present
status of our understanding of the factors governing the rates
of hydrolysis of pesticides which are sorbed to sediments. The
work reported herein deals specifically with abiotic hydrolysis
reactions, which for some pesticides, may be as important or
more important than biologically mediated hydrolysis reactions
(7, 10-13).

Until recently, no efforts to measure the rates of
hydrolytic degradation of sorbed pesticides have been
reported. Indeed, it has been widely assumed that hydrolytic
reactions are important only in the aqueous phase and that
hydrolysis of sorbed pesticides proceeds at an insignificant
rate (13).

The only available evidence in the literature which
relates, though indirectly, to hydrolysis of sorbed pesticides
concerns pesticides in soil systems (see for example 14, 15).
Though the results of such studies are not directly applicable
to aquatic systems, they do, in general, show that certain
pesticides undergo abiotic reactions in soil-sorbed states.

This review, then, reports results of experiments which
provide information that can be used to test the hypothesis that
hydrolysis reactions proceed at substantially reduced rates when
the molecules undergoing hydrolysis are sorbed to sediments.
Results are reported for a variety of pesticides and for model
compounds that are similar in structural features to
pesticides. Included are neutral, base-catalyzed and, to a
limited extent, acid-catalyzed hydrolysis reactions.

Preliminary Considerations

Three general classes of hydrolytic reactions in aqueous
solutions have been characterized. In neutral, or pH
independent hydrolysis, the rate of disappearance of a
pesticide, P, is given by

$$\frac{d[P]}{dt} = - k_1 \, [P] \tag{1}$$

where k_1 is the first-order disappearance rate constant. For
base-mediated hydrolysis, the corresponding expression is

$$\frac{d[P]}{dt} = -k_B[B][P] = -k_{obs}[P] \tag{2}$$

where B represents a generalized base. For natural waters and
the experimental systems relevant to this report, the only base
of significance for such reactions is the hydroxide ion, OH^-
(16). In equation 2, k_{obs} represents a pseudo first-order rate
constant, valid at fixed pH (or [B]).

Recently reported results for the hydrolysis kinetics of chlorpyrifos (7) suggest that equation 2 may not be a valid representation of alkaline hydrolysis kinetics for at least one class of pesticides (organophosphorothioates). In short, k_B may be pH dependent. However, disappearance kinetics for such molecules are still adequately described at fixed pH by pseudo first-order kinetics.

Acid-catalyzed hydrolysis kinetics are described by the expression

$$\frac{d[P]}{dt} = -k_a[H^+][P] = -k_{obs}[P] \tag{3}$$

where $[H^+]$ represents the hydrogen ion activity, and k_{obs} the pseudo first-order disappearance rate constant at fixed pH.

For a given pesticide which undergoes hydrolysis, any or all of these hydrolytic pathways may be relevant at various pH's. Organophosphorothioates, for example, have measurable neutral and alkaline hydrolysis rate constants (7). Esters of 2,4-dichlorophenoxyacetic acid (2,4-D), on the other hand, hydrolyze by acid and alkaline catalyzed reactions, but have extremely small neutral hydrolysis rate constants (17). Thus, any study of the hydrolysis of sorbed pesticides must be prefaced by an understanding of the hydrolytic behavior of individual pesticides in aqueous solution.

Another important consideration in investigation of the reaction of sorbed pesticides is the nature of the sorption process itself. Sorption/desorption kinetics and the physicochemical characteristics of the pesticide molecules in the sediment-sorbed state can be expected to influence the kinetic observations made in experimental systems.

Sorption has been commonly described as an equilibrium process, in which the pesticide molecules are rapidly and readily exchanged between the sediment and aqueous phases. In this approach (8), the equilibrium water phase concentration, C_w (expressed relative to suspension volume) is related to the sediment phase concentration, C_s (expressed relative to dry weight sediment), through

$$K_p = C_s/C_w \tag{4}$$

where K_p is the equilibrium partition coefficient $(L-g^{-1})$ for the pesticide. C_w is then related to the total concentration of pesticide, C_T, by

$$C_w = \frac{C_T}{1+\rho K_p} \tag{5}$$

where ρ is the sediment-to-water ratio. K_p has been shown to be directly proportional to the weight fraction of organic carbon in the sediments (O.C.)

$$K_p = (O.C.) \, K_{oc} \qquad\qquad (6)$$

For a given pollutant, K_{oc} has been shown in turn to be directly related to the octanol-water partition coefficient of the pesticide (18).

Several investigations have, however, verified the inadequacy of this representation of the sorption process. Variations of "K_p" with sediment concentration (19, 20) have been reported. More importantly, the rate of the sorption process has been shown to be more complex than a simple rapid equilibrium between sediment and aqueous phases (9, 10, 21).

The fact that sorptive equilibrium can be approached quite slowly is illustrated dramatically by data for the system in which chlorpyrifos is sorbed to EPA-14, one of a group of sediments collected and characterized for the U. S. Environmental Protection Agency (22). Figure 1 is a plot of the sediment/aqueous concentration ratio versus time for this system. It is characterized by a rapid sorption process and a much slower sorption process which does not reach equilibrium until about 10 days after initial mixing of the sediment and chlorpyrifos solution.

Though this system is perhaps an extreme example of slow sorption kinetics, it illustrates that the assumption of rapid equilibrium between the sediment and aqueous phases is questionable. The importance of such an observation to the investigation of hydrolysis kinetics in sediment/water systems must be emphasized. Certainly, any model of hydrolysis kinetics in sediment/water systems must include explicit expressions for the kinetics of the sorption/desorption process.

Unfortunately, our present understanding of sorption kinetics is inadequate to allow unambiguous representation of the sorption-desorption process. Clearly the states of sorbed pesticides include fractions which vary in their lability with respect to desorption (9, 10, 21). The fraction of the sorbed molecules in relatively labile and non-labile states is a function of the nature of the pesticide and sediment and the time of contact between the sediment and pesticide solution.

With these limitations in mind, however, we have used the following model to represent the kinetics of the hydrolysis of pesticides in sediment/water suspensions (10)

$$
\begin{array}{ccc}
 & k_I & \\
C_w & & C_s \qquad\qquad (7) \\
 & k_o & \\
k_w & & k_s \\
\end{array}
$$

$$\text{Products}$$

Here, k_I represents a pseudo first-order rate constant (linear in sediment concentration) for the sorption process, k_o the

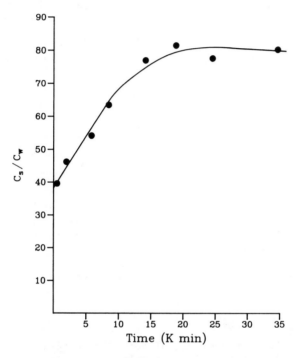

Figure 1. Kinetics of the sorption of chlorpyrifos to EPA-14 sediment, ρ = 0.20, t = 25 ^{O}C.

first order rate constant for desorption, k_w the first or pseudo first-order rate constant for hydrolysis of the aqueous (dissolved) pesticide and k_s the corresponding rate constant for hydrolysis of the sorbed pesticide.

This model, in light of the discussion above, is clearly not representative of all of the kinetic processes which are occurring in sediment/water systems containing hydrophobic pesticides. However, it does include at least the more labile fraction of the sorbed pesticide in the overall kinetic model. Complications due to the inadequacy of this representation will be illustrated and discussed below.

Based on this model, the following rate equations relating the hydrolytic degradation of pesticides from sediment water suspensions can be written:

$$\frac{dC_w}{dt} = -(k_w + k_I)C_w + k_o C_s$$

$$\frac{dC_s}{dt} = -(k_s + k_o)C_s + k_I C_w \qquad (8)$$

$$\frac{dPr}{dt} = k_s C_s + k_w C_w$$

where Pr represents the concentration of products in the aqueous phase. Products are uniformly more hydrophilic than the parent pesticide molecules, making sorption of product molecules a relatively unimportant consideration.

To the extent that overall disappearance kinetics in such systems are pseudo first order, we may also write

$$\frac{dC_T}{dt} = -k_{obs} C_T \qquad (9)$$

Note that if hydrolysis of sorbed molecules is unimportant and sorption/desorption is fast relative to hydrolysis, then, from equations 5 and 9,

$$k_{obs} = k_w / (1 + \rho K_p) \qquad (10)$$

Experimental Considerations

Details concerning the experimental procedures used to obtain the data upon which the results discussed below are based have been presented elsewhere (7, 10-12, 17). Three types of experiments are involved.

First, determination of hydrolysis kinetics for each compound in sediment-free distilled, buffered distilled or natural water systems were measured. Using sterile techniques, concentrations of the parent compounds were determined as a

function of time. Generally, these experiments were conducted
by withdrawing aliquots of an aqueous solution of the compound
from the temperature-controlled reaction flask, extracting the
parent compound from the aliquot using an organic solvent
(usually iso-octane) and determining concentrations of the
parent compound using gas-liquid chromatography.

Such experiments were repeated for each compound at a
variety of pH's and temperatures so that pH-rate constant
profiles and activation energies could be obtained. Extraneous
experimental complications such as sorption of the compound to
container walls, incomplete extraction from aqueous solutions
and possible catalysis by metal ions in solution were carefully
monitored and accounted for in the final determination of
aqueous phase hydrolysis rate constants. Of these
possibilities, only sorption to container walls was observed to
have a measurable effect on the experimental data.

In addition, most of these aqueous phase experiments
included product identification using gas chromatographic-mass
spectrometric (GC-MS) or liquid chromatographic-MS techniques.
Product analyses were used to verify that disappearance kinetics
were indeed due to hydrolysis reactions.

In a second class of experiments, detailed studies of
disappearance kinetics in sediment-water systems were
performed. A series of centrifuge tubes was assembled
containing identical concentrations of parent compound and
identical sediment-to-water ratios.

The tubes were gently agitated at constant temperature and
sacrificed at regular time intervals. Each tube was centrifuged
and the bulk of the water phase separated from the sediment
phase. The aqueous and sediment phases were then separately
analyzed for the remaining concentrations of parent compound.
Careful tests were conducted to assure complete extraction of
the compound from the two phases (10). Thus, the concentrations
remaining in each phase could be separately determined as a
function of time.

The third class of experiments was similar to the second,
except that the centrifugation step was omitted and only the
total parent compound concentration in the sediment-water
systems was determined as a function of time.

Finally, several types of auxilliary experiments were
performed during various phases of these investigations.
Included were purity checks on parent compounds, determinations
of K_p values for selected compounds on various sediments,
investigations of the effect of pH on measured K_p's,
verification of the validity of pH measurements in dilute
sediment-water slurries, elimination of the possibility of
buffer catalysis of hydrolysis reactions in our systems, and
qualitative product analyses from sediment/water hydrolysis
studies.

All sediments used in the studies reported in this article
were selected from a group of well-characterized soil and
sediment samples which were collected and air-dried for the U.
S. Environmental Protection Agency by Hassett and co-workers
(22). The characteristics of several of these are outlined in
Table I.

Table I.   Characteristics of Sediments Used in These Studies

	Sediment				
	EPA-12	EPA-13	EPA-14	EPA-23	EPA-26
pH (1:2)	7.55	6.74	4.3	7.1	8.1
CEC (meq/100g)	13.53	11.9	18.9	31.15	20.86
total N (%)	0.214	0.167	0.064	0.195	0.152
organic carbon (%)	2.33	3.04	0.48	2.38	1.48
sand/silt/ clay (%)	0/35/64	20/53/27	2/64/34	17/69/14	2/43/55
K_p for chlopyrifos: calculated (from K_{ow})[a]	465	608	96	475	300
measured (± S.D.)	–	–	83 ± 6	403 ± 92	307 ± 96

[a]Reference 22

Results and Discussion

Neutral Hydrolysis Studies. Investigations of neutral (pH-
independent) hydrolysis kinetics in sediment/water systems were
conducted for three organophosphorothioate insecticides
(chlorpyrifos, diazinon and Ronnel), 4-(p-chlorophenoxy)butyl
bromide, benzyl chloride, and hexachlorocyclopentadiene.
    1. Chlorpyrifos, O-O-diethyl O-(3,5,6-trichloro-2-pyridyl)
phosphorothioate, is the compound for which the most exhaustive
kinetic investigations were conducted (10). The kinetics of the
hydrolysis as a function of pH in distilled and buffered
distilled water systems is summarized by the pH-rate profile
shown in Figure 2 (7). The value, k_{obs}=(6.22±0.09) x 10^{-6} min^{-1}
is the neutral hydrolysis rate constant for chlorpyrifos in
distilled water at 25°C.
    Values at 25°C for the hydrolysis rate constants of
chlorpyrifos in sterilized natural waters and in water isolated
from a 25 g/1 slurry of EPA-14 in distilled water (stirred for 1
week prior to separation of the sediment and water phases) are

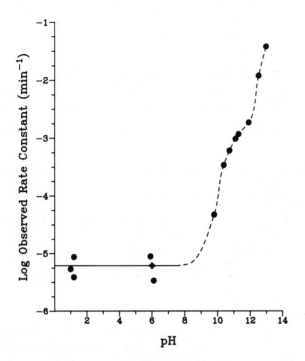

Figure 2. pH-rate constant profile for chlorpyrifos in distilled water at 25 °C.

shown in Table II. The variability in these data and the
apparent increase in the rate constants from distilled to
natural waters are unexplained. Biological, trace metal and
glass-surface catalysis have been eliminated as possible sources
of this variability and rate enhancement. Biological
degradation was determined (7) to be unimportant for
chlorpyrifos in our systems, so sterility is not a factor. A
value of k_{obs}=1.0 x 10^{-5} min^{-1} was selected as a reasonable
approximation of the hydrolysis rate constant in the aqueous
phases of the sediment-water systems used in these studies.

Table II. Ranges of Measured Rate Constants for the Disappear-
           ance of Chlorpyrifos from Natural Water Samples at
           25°C

Samples	Rate Constants[a]
Oconee River water (20)[b]	(0.71 – 4.00) x 10^{-5} min^{-1}
Hickory Hills Pond (12)[b]	(1.00 – 3.90) x 10^{-5} min^{-1}
EPA – 14 supernatant (6)[b]	(0.64 – 0.97) x 10^{-5} min^{-1}

[a]Ranges shown represent a composite of autoclaved and filter
sterilized trials, along with trials involving non-sterilized
waters. No bacterial effects are in evidence. Effects of pH
are unimportant in these waters. Metal catalysis is not a
factor. For a detailed analysis of the data including error
analysis, see reference 7.
[b]Number of kinetic runs in parentheses.

The data from a representative study of the disappearance
of chlorpyrifos from an EPA-14 sediment/water system (ρ=0.20,
fraction sorbed = 0.94) is illustrated in Figure 3. Comparison
with Figure 1 shows that once sorptive equilibrium is achieved
(t>14,000 minutes) the disappearance rate is first order for
both the water and sediment phases. Also, the aqueous
disappearance rate constant calculated from the slope of the
linear portion of the natural log aqueous concentration versus
time plot is 0.5±0.2 x 10^{-5} min^{-1}, which is similar to the
values measured in sediment-free EPA-14 supernatant (Table
II). A plot summarizing two experiments using EPA-23 sediment
is shown in Figure 4. The value of k_w calculated from the
natural log water concentration vs. time plot in this figure is
(1.9±0.2) x 10^{-5} min^{-1}.
Data from these studies were analyzed by a computer using
equations 8 based on our simple kinetic model for the
sediment/water systems (eqn. 7). The computer program (23) uses
concentrations of chlorpyrifos in the water and sediment phases
and product concentrations (obtained by difference) as a

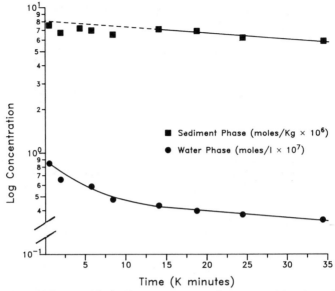

Figure 3. Chlorpyrifos disappearance from an EPA-14 sediment/water system, ρ= 0.20, t = 25 ^{o}C.

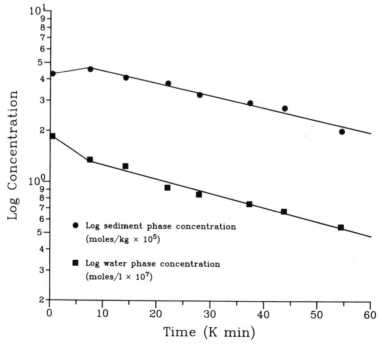

Figure 4. Chlorpyrifos disappearance from EPA-23 sediment/water systems, ρ= 0.016, t = 25 ^{o}C.

function of time to calculate values for any three of the rate constants k_I, k_o, k_w and k_s. For the purposes of our calculations, a "known" value of 1.0×10^{-5} min^{-1} for k_w was used to enable calculation of k_I, k_o and k_s. The results of these calculations are shown in Table III. Also shown in Table III are values for k_{obs}, the overall chlorpyrifos disappearance rate constant, and a value calculated for k_{obs} using equation 10, which is based on $k_s = 0$, i.e. no hydrolysis of sorbed chlorpyrifos.

Table III. Observed and Calculated Values of Rate Constants (min^{-1}) for Chlorpyrifos in Sediment/Water Systems at Non-adjusted pH's.[a]

	EPA 14	EPA 23	EPA 23
sediment/water (ρ) fraction	0.20	0.016	0.016
sorbed	0.94	0.87	0.87
sterile?	yes	no	yes
k_{obs}	1.0×10^{-5}	1.7×10^{-5}	1.6×10^{-5}
k_{obs} (calc.)[b]	6.0×10^{-7}	1.3×10^{-6}	1.3×10^{-6}
k_I	$6 \pm 2) \times 10^{-3}$	$(4 \pm 1) \times 10^{-3}$	$(4 \pm 1) \times 10^{-3}$
k_o	$(6 \pm 2) \times 10^{-4}$	$(9 \pm 3) \times 10^{-4}$	$(4 \pm 2) \times 10^{-4}$
k_w (fixed)[c]	1.0×10^{-5}	1.0×10^{-5}	1.0×10^{-5}
k_s	$(6.9 \pm 0.9) \times 10^{-6}$	$(1.8 \pm 0.1) \times 10^{-5}$	$(1.2 \pm 0.1) \times 10^{-5}$
pH (H_2O phase)	4.1 ± 0.4	7.2 ± 0.2	7.4 ± 0.4

[a](See text for Symbol Definitions).
[b]Assuming sediment/water equilibrium, no hydrolysis in the sorbed state, and $k_w = 1.0 \times 10^{-5}$, $k_{obs} = k_w/(1 + \rho K_p)$.
[c]For computer model calculations.

Several features of these calculations are important. First, the computer calculated uncertainties shown for the calculated values of k_I, k_o and k_s are an indication that the model has considerable validity for describing the kinetics of the system, at least over one half-life in the disappearance of chlorpyrifos. Second, the values of k_o and k_I are all similar and their magnitude indicates that in this case the assumption of rapid sorption/desorption kinetics compared to hydrolysis is valid.
More importantly, the calculated values of k_s are all similar in magnitude to k_w. Coupled with the fact that the

values calculated for k_{obs} assuming no sediment phase hydrolysis
are all considerably lower than the actual values for k_{obs},
these k_s values indicate that <u>neutral hydrolysis of chlorpyrifos</u>
<u>in the sorbed state proceeds at a rate that is the same as the</u>
<u>disappearance rate in the aqueous phase.</u>
    2. Diazinon and Ronnel. The conclusion that neutral
hydrolysis of sorbed chlorpyrifos is characterized by a first-
order rate constant similar to the aqueous phase value is
strengthened and made more general by the results for diazinon,
0,0-diethyl 0-(2-<u>iso</u>-propyl-4-methyl-6-pyrimidyl)
phosphorothioate, and Ronnel, 0,0-dimethyl 0-(2,4,5-
trichlorophenyl) phosphorothioate (10). The results for the pH
independent hydrolysis at 35°C for these compounds in an EPA-26
sediment/water system (ρ=0.040) are summarized in Table IV.
Because the aqueous (distilled) values of k_w for diazinon and
Ronnel are similar in magnitude to the value for chlorpyrifos,
and because these values were shown by the chlorpyrifos study to
be slow compared to sorption/desorption kinetics, computer
calculations of k_s were not deemed necessary and were not made
for these data.

Table IV.   Experimental Values for Neutral Hydrolysis
            Disappearance Rate Constants (min^{-1}) in an EPA 26
            Sediment/Water System for Diazinon and Ronnel at
            $35°C^a$

	Diazinon	Ronnel
sediment/water(ρ)	0.040	0.040
fraction sorbed	0.64	0.96
k_{obs}	$(3 \pm 1)x10^{-5}$	$(2.7 \pm 0.4)x10^{-5}$
k_{obs} (calculated)^b	$1.3x10^{-5}$	$2.3x10^{-6}$
k_w (distilled)	$(1.2 \pm 0.2)x10^{-5}$	$(2.0 \pm 0.2)x10^{-5}$
k_w (observed)	$(3.8 \pm 0.6)x10^{-5}$	$(3.8 \pm 1.8)x10^{-5}$
k_s	$(2.9 \pm 0.5)x10^{-5}$	$(2.6 \pm 0.3)x10^{-5}$

^{a}Symbols Defined in text.
^{b}Assuming sediment/water equilibrium, no hydrolysis in the
sorbed state and k_w = 3 x k_w (distilled)-see discussion. k_{obs} =
$k_w/(1 + \rho K_p)$.

    Again the values of k_s, calculated from the log C_s vs. time
plots, were similar in magnitude to the value of k_w calculated
from the log C_w vs. time plots. Also the values of k_{obs}
calculated from the k_s=0 assumption implicit in equation 10 were
lower than the experimental k_{obs} values. This latter effect is
less dramatic for diazinon since its lower K_p value results in
an equilibrium fraction sorbed of only 0.64. Note also that the

values for k_w are 2-3 times the distilled water values. This observation is also consistent with the rate enhancements observed for chlorpyrifos in natural (cf. distilled) waters.

Thus, for chlorpyrifos, diazinon, Ronnel (and by extension, other organophosphorothioate pesticides), neutral hydrolysis proceeds at similar rates in both the aqueous and sediment phases of sediment/water systems.

3. Experiments on the hydrolysis of 4-(p-chlorophenoxy) butyl bromide, (PCBB) which proceeds via an S_N2 substitution mechanism (11) were similar in design and data analysis procedures to the chlorpyrifos experiments detailed above. Results from a study at 35°C using EPA-12 sediment with 80% of the compound in the sorbed state are illustrated in Figure 5. Calculated and observed values from this study, using the distilled water value for k_w of $(7.9\pm0.5\text{x}10^{-5})$ min^{-1} as a "known" value for computer calculations are:

$$k_I = 8.1 \times 10^{-5} \text{ min}^{-1}$$
$$k_O = 1.6 \times 10^{-5} \text{ min}^{-1}$$
$$k_S = 5.1 \times 10^{-5} \text{ min}^{-1}$$

Again, the value of k_S is similar in magnitude to the value of k_w. Other studies using EPA-12 (80-95% sorbed) at 25°C and EPA-10 (90% sorbed) at 35° also indicate similar values for k_S and k_w.

Several features of the PCBB experiments are different than those for chlorpyrifos. The hydrolysis reaction proceeds via a different mechanism. The rate enhancements observed for chlorpyrifos in natural waters and the aqueous phases of the sediment/water systems (as compared to sterile distilled water) are not observed for PCBB. The values of k_I and k_O calculated for PCBB are slower than those for chlorpyrifos and similar in magnitude to the hydrolysis rates.

In spite of these differences, neutral hydrolysis is still characterized by similar rate constants for both sediment-sorbed and aqueous PCBB.

4. Benzyl chloride hydrolysis proceeds via a third mechanism (S_N1). Results of studies of benzyl chloride hydrolysis (11) in distilled water and EPA-13 and EPA-2 sediment/water systems are summarized in Table V. Results for this compound include only overall first-order disappearance rate constants, but the data clearly show that the hydrolysis rate is independent of the fraction sorbed to sediment. Thus, the conclusion is again made that neutral hydrolysis proceeds via similar rate constants in both the aqueous and sediment-sorbed phases.

Table V. Hydrolysis of Benzyl Chloride in Sediment/Water
Systems at 25°C

k_{obs}, min^{-1}x10^3	Sediment to Water Ratio, ρ(Sed. #)	Fraction Sorbed
1.18 ± 0.05	0	0
1.33 ± 0.03	0	0
1.1 ± 0.1	0.025 (EPA 13)	0.15
1.4 ± 0.1	0.05 (EPA 13)	0.25
1.15 ± .05	1.0 (EPA 2)	0.87
1.10 ± .08	1.0 (EPA 2)	0.87

5. The hydrolysis of hexachlorocyclopentadiene (HEX)
represents a fourth hydrolysis mechanism (S_N2'). Studies of the
overall disappearance kinetics of HEX from sterile distilled
water and EPA-13 sediment/water systems (12) are summarized in
Table VI. Again, the rate constants are essentially independent
of the sediment concentration and therefore independent of the
fraction of the HEX which is sorbed to the sediment. This
indicates that neutral hydrolysis of HEX is also characterized
by similar rate constants for both the sediment-sorbed and
aqueous phases.

Table VI. Hydrolysis of Hexachlorocyctopentadiene in EPA-13
Sediment/Water Systems at 30°C

Sediment to Water Ratio, ρ	k_{obs}, min^{-1} x 10^5
0	9 ± 3
0.05	21 ± 2
0.10	26 ± 2
0.15	27 ± 4
0.20	22 ± 2
0.40	16 ± 1
1.0	20 ± 2
2.0	13 ± 2

In summary, neutral hydrolysis rate constants for six
different compounds which hydrolyze by four different hydrolysis
mechanisms were determined in sediment/water systems using seven

different sediments. Sediment to water ratios varying from
5×10^{-4} to 2.0 were used. Yet, in each case, neutral hydrolysis
of the sorbed compounds was shown to be characterized by rate
constants which were very nearly equal to the rate constants for
hydrolysis in the aqueous phase of these systems. The
conclusion is that neutral hydrolysis reactions are not quenched
when the molecules undergoing hydrolysis are sorbed to
sediments. In fact the rate of neutral hydrolytic reactions
appears to be unaffected by sorption.

Alkaline Hydrolysis Studies. Alkaline catalyzed hydrolysis
kinetics in sediment/water systems have been investigated for
chlorpyrifos and the methyl and n-octyl esters of 2,4-
dichlorophenoxyacetic acid (2,4-D).
    1. Chlorpyrifos. As was the case for the neutral
hydrolysis studies, the most detailed kinetic investigations of
alkaline hydrolysis kinetics in sediment/water systems have been
conducted using chlorpyrifos (10). As can be seen from Figure
2, alkaline hydrolysis of chlorpyrifos is not second-order, so
the value selected for k_w cannot be calculated from the pH and a
second-order rate constant. Nevertheless, since aqueous
kinetics at alkaline pH's for chlorpyrifos was always pseudo-
first order, careful pH measurements and Figure 2 can be used to
select accurate values for k_w at any pH.
    A preliminary consideration for studies at alkaline pH's is
the effect of pH on K_p values and on sorption/desorption rate
constants. Studies using chlorpyrifos and EPA-26 ($\rho = 0.0150$)
indicate no measurable effects on K_p over the pH range 5.5-10.8
($K_p = 250 \pm 37$ for five determinations). Kinetic effects are also
minor, as illustrated by the similarity in the calculated values
of k_T and k_o for EPA-23 at pH's of 7.2 and 7.4 (Table III) and
10.67 (Table VII).
    Two types of investigations of the alkaline hydrolysis of
chlorpyrifos in sediment/water systems were made, all at pH's
between 10.6 and 10.8. First, studies were conducted in which
the pH was adjusted (using a carbonate buffer) immediately upon
mixing the sediments (EPA-23 and EPA-26) with the chlorpyrifos
solution. Second, a study using EPA-26 was made in which the
alkaline buffer was not added until three days after mixing the
sediment with the chlorpyrifos solution. Three days represents
a time which is long with respect to the achievement of
sediment-water equilibrium for this system, yet short compared
to the neutral hydrolysis half life (~50 days).

Table VII.   Experimental and Calculated Values of the Rate
Constants for the Alkaline Hydrolysis of
Chlorpyrifos in Sediment/Water Systems[a]

	EPA 23	EPA 26
pH	10.67 ± 0.04	10.60 ± 0.04
ρ	0.019 ± 0.001	0.031 ± 0.001
K_p	453 ± 59	191 ± 4
ave fr. sorbed	0.90	0.85
k_{obs}	$(1.05 \pm 0.11) \times 10^{-4}$	$(1.10 \pm 0.04) \times 10^{-4}$
k_{obs}(calculated)[b]	5.0×10^{-5}	6.2×10^{-5}
k_I	$(1.3 \pm 0.8) \times 10^{-3}$	$(2.5 \pm 0.9) \times 10^{-2}$
k_o	$(1.8 \pm 1.1) \times 10^{-4}$	$(4.4 \pm 1.7) \times 10^{-3}$
k_w(fixed)[c]	4.8×10^{-4}	4.3×10^{-4}
k_s	$(7.1 \pm 0.8) \times 10^{-5}$	$(4.1 \pm 0.3) \times 10^{-5}$

[a]Symbols defined in the text. No pre-equilibrium between
sediment and chlorpyrifos prior to pH adjustment.
[b]Assuming sediment/water equilibrium, no hydrolysis in the
sorbed state, k_{obs}(calc.) $= k_w/(1+\rho K_p)$.
[c]For computer calculations. The value of k_w is the expected
distilled water hydrolysis rate constant at this pH.

In the first type of study, pseudo first-order kinetics
were observed in both the sediment and aqueous phases from t=0
through two half-lives in overall chlorpyrifos disappearance
(total time ~8 days). For these studies, computer calculations
using the model illustrated in equations 7 were again used to
calculate values for k_I, k_o and k_s, assuming a value of k_w equal
to the pseudo first-order rate constant in distilled water
buffered to the same pH. Values were also calculated for k_{obs}
assuming $k_s=0$ (equation 10) for comparison to the experimental
k_s values. The results of these calculations are shown in Table
VII.

The contrast between these alkaline hydrolysis results and
the neutral hydrolysis results is striking. The calculated
values of k_s are lower by factors of 7-10 than the corresponding
k_w values. The values calculated for k_{obs} assuming $k_s=0$ are
only 1.8-2.1 times smaller than the experimental k_{obs} values
(cf. calculated values 12-17 times lower than observed for
neutral hydrolysis at similar fractions sorbed).

These results, therefore, show that alkaline hydrolysis is
considerably slowed when the chlorpyrifos is sorbed to
sediments.

The results from the study featuring sediment-chlorpyrifos
equilibration prior to pH adjustment (Figure 6) are

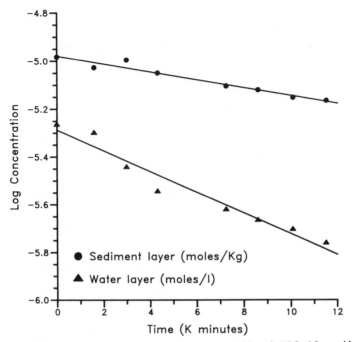

Figure 5. PCBB disappearance from a sterilized EPA-12 sediment/
water system, ρ = 0.050, t = 35 °C.

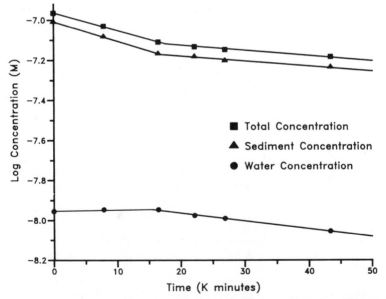

Figure 6. Chlorpyrifos disappearance from an EPA-26 sediment/
water system equilibrated for three days prior to pH adjustment
to 10.6; ρ = 0.031, t = 25 °C.

qualitatively different than those from the studies without pre-
equilibration. Attempts to fit these data to our simple kinetic
model gave a very poor fit. The model is clearly inadequate in
this case. Though this is a very limited data set, the kinetics
appear to follow an initial disappearance of chlorpyrifos
dominated by alkaline hydrolysis of the aqueous phase material
and desorption of a fraction of the sorbed material. Through
approximately one half-life, k_{obs} is 2×10^{-4} min^{-1}, a value
similar to the k_{obs} measured for the EPA-26 study without pre-
equilibration (1×10^{-4} min^{-1}). Subsequently, however, k_{obs} falls
to 5×10^{-5} min^{-1}, a number quite similar to the k_s value
calculated for the parallel study without pre-equilibration
(4×10^{-5} min^{-1}).

   These observations, though tentative, suggest the existence
of a substantial fraction of the sorbed material which is
considerably less labile with respect to desorption than the
material initially sorbed to the sediments. Further study of
these effects is clearly needed.

   2. Esters of 2,4-D. Studies of the alkaline hydrolysis of
the methyl and n-octyl esters of 2,4-D in sediment/water systems
(24), though less detailed than the chlorpyrifos studies, show
similar effects. Results from investigations using EPA-13 at
pH's near 10 for the methyl and octyl esters of 2,4-D are
summarized in Figure 7. Under the conditons in these
experiments, the fractions of the methyl and octyl esters which
are sorbed to the sediment are 0.10 and 0.87, respectively. The
aqueous hydrolysis half-lives of the methyl and octyl esters at
pH=10 are 3.6 and 27 minutes, respectively. In the
sediment/water system, the methyl ester, which is mainly in the
dissolved phase, hydrolyzes at a rate similar to that expected
for the sediment-free system at the same pH. The octyl ester,
on the other hand, hydrolyses at a rate which is considerably
retarded (and non-first-order) when compared to the expected
aqueous phase rate. Though the data are less detailed and do
not permit calculations similar to those conducted for
chlorpyrifos, it is clear that the effect of sorption is to
considerably slow the alkaline hydrolysis rate.

   Studies of the disappearance of the octyl ester at pH 9.8
in sediment/water systems aged 3 days prior to pH adjustment are
summarized in Figure 8. For the systems with $\rho=0.013$ and 0.005
(fractions sorbed = .978 and .945) the rate is pseudo first
order, but the rate constant is 10^4 times smaller than the
aqueous value (1.6×10^{-1} min^{-1}) at this pH. As was suggested for
chlorpyrifos, this k_{obs} value may be characteristic of the
actual value of k_s. At $\rho=0.001$, (fraction sorbed = 0.78), the
disappearance kinetics is not first order, but shows rapid
disappearance of the aqueous ester, followed by disappearance of
the sorbed ester at a rate similar to the studies with higher
sediment to water ratios.

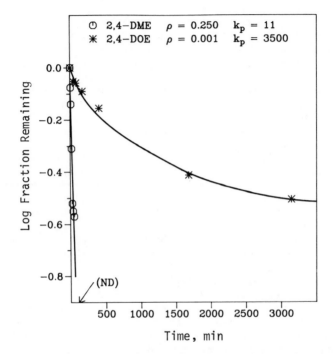

Figure 7. Alkaline hydrolysis of the 2,4-D methyl ester (pH 10.01) and 2,4-D n-octyl ester (pH 10.12) from heat-sterilized EPA-13 sediment/water systems.

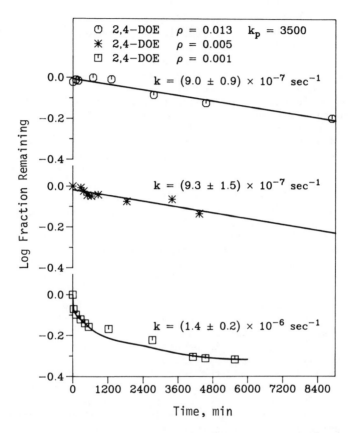

Figure 8.   Alkaline hydrolysis of the 2,4-D n-octyl ester (pH 9.8) in EPA-13 sediment/water systems equilibrated 3 days prior to pH adjustment.

Based on experiments for these three compounds, it is clear
that the alkaline hydrolysis of sorbed molecules is
substantially retarded with respect to the rate for dissolved
molecules. The extent of this retardation cannot be
quantitatively discussed at this time, however, due to lack of a
sufficiently broad set of detailed experimental data.

3. Acid Hydrolysis. Considering the observed results for
neutral and alkaline hydrolysis, it is interesting to speculate
as to the expected effects of sorption on acid-catalyzed
hydrolysis. Since most sediments are of predominantly clay
mineralogy, sediment grains exhibit generally negative surface
changes at pH's common in natural waters (25). When this is the
case, one would expect the concentration of negative ions such
as OH^- to be lower near the sediment particle surface. In
addition, the (acidic) functional groups of the organic matter
associated with the sediments would be expected to be negatively
charged at alkaline pH's. These effects alone might be expected
to reduce the rate of alkaline hydrolytic processes occuring at
these surfaces.

Also, recall that alkaline hydrolysis of
organophosphorothioates has been shown to involve a negatively-
charged intermediate (v.s. and 7). Such an intermediate would
be expected to be less stable in the negatively charged
environment of the sediment particle surface.

On the other hand, the concentration of positive ions,
including H_3O^+, near sediment particle surfaces would be
expected to be enhanced relative to the bulk solution
concentrations. From this consideration, we would predict that
acid-catalyzed hydrolysis reactions should occur at enhanced
rates for sorbed molecules.

Unfortunately, there is presently only a very tentative bit
of evidence available to substantiate this prediction. In one
experiment at an acid pH, the overall rate of hydrolysis of the
n-octyl ester of 2,4-D was measured in a sediment/water slurry
in which a substantial fraction of the ester was sorbed. The
rate was observed to be substantially faster than the predicted
aqueous phase rate. Though this is an indication that the above
prediction is correct, much more experimental work is needed to
substantiate and quantify this predicted rate enhancement.

Summary and Conclusions

The hypothesis that hydrolysis of sorbed molecules occurs at
rates insignificant with respect to aqueous phase hydrolysis has
been demonstrated to be incorrect for neutral (pH-independent)
hydrolysis reactions. The rate-constants for sorbed state
neutral hydrolysis are, on the contrary, similar in magnitude to
those for hydrolysis in the aqueous phase.

The hypothesis has been shown to be more nearly correct for
alkaline hydrolysis reactions, since significant rate

retardations, corresponding to greatly reduced rate constants for sorbed molecules, occur when substantial fractions of the hydrolyzing molecules are sorbed to sediments.

Inadequate understanding of the kinetics of the sorption/desorption process detracts from our ability to completely understand the effects of sorption on hydrolytic rates, and more research is needed in this regard.

Limited experimental evidence and theoretical considerations suggest that acid-catalyzed hydrolysis rates are subsantially enhanced for sorbed molecules. Much more experimental evidence is necessary, however, to verify this effect.

These conclusions have several implications for pesticide waste disposal considerations. For incidental or accidental disposal of pesticides in natural aquatic systems, the results suggest that model calculations using aqueous solution values for abiotic neutral hydrolysis rate constants can be used without regard to sorption to sediments. For alkaline hydrolysis, on the other hand, models must explicitly include sorption phenomena and the corresponding rate reductions in order to accurately predict hydrolytic degadation.

The limited acid-hydrolysis results, if substantiated, have broader implications. They suggest that rapid hydrolysis of pesticide wastes in acidified artificial sediment/water slurries may be an attractive method for the intentional disposal and degradation of pesticide wastes.

Acknowledgments

Most of the work reviewed in this article was performed at the U. S. Environmental Research Laboratory in Athens, GA. Support for Donald Macalady's work at this laboratory was provided by a Senior Associateship award from the National Rsearch Council. NOTE: Mention of trade names or commercial products does not constitute endorsement or recommendation for use by the U. S. Environmental Protection Agency.

Literature Cited

1. Wolfe, N.L.; Zepp, R.G.; Paris, D.F. Water Research, 1978, 12, 561.
2. Wolfe, N.L.; Burns, L.A.; Steen, W.C. Chemosphere, 1980, 9, 393.
3. Wolfe, N.L.; Zepp, R.G.; Doster, J.C.; Hollis, R.C. J. Agric. Food Chem., 1976, 24, 1041.
4. Wolfe, N.L.; Zepp, R.G.; Paris, D.F. Water Reserch, 1978, 12, 565.
5. Maquire, R.J.; Hale, E.J. J. Agric. Food Chem., 1980, 28, 372.
6. Ou, L.T.; Gancarz, D.H.; Wheeler, W.B.; Rao, P.S.C.; Davidson, J.M. J. Environ. Quality, 1982, 11, 293.

7. Macalady, D.L.; Wolfe, N.L., J. Agric. Food Chem., 1983, 31, 1139.
8. Karickhoff, S.W.; Brown, D.S.; Scott, T.A. Water Research, 1979, 13, 241.
9. Karickhoff, S.W. In Contaminants and Sediments, Baker, R.A., Ed. Ann Arbor Science, Ann Arbor, MI, 1980; Vol. 2, Chapter II.
10. Macalady, D.L.; Wolfe, N.L. "Effects of Sediment Sorption on Abiotic Hydrolyses: I. Organophosphorothioate Esters", submitted to Environ. Sci. Technol., 1983.
11. Pierce, J.H.; Wolfe, N.L. "Effects of Sediment Sorption on Abiotic Hydrolyses: II. 4-(p-chlorophenoxy) butyl bromide and Benzyl Chloride", paper in preparation, 1984.
12. Wolfe, N.L.; Zepp, R.G.; Schlotzhauer, P.; Sink, M. Chemosphere 1982, 11, 91.
13. Wolfe, N.L.; Zepp, R.G.; Paris, D.F.; Baughman, G.L.; Hollis, R.C. Environ. Sci. Technol. 1977, 11, 1077.
14. Sethunan, N.; MacRae, I.C., J. Agric. Food Chem. 1969, 17, 221.
15. Mingelgrin, U.; Saltzman, S.; Yaron, B. Soil Sci. Soc. Am. J. 1977, 41, 519.
16. Perdue, E.M.; Wolfe, N.L. Environ. Sci. Technol. 1983, 11, 635.
17. Wolfe, N.L., U.S. Environmental Protection Agency, Athens, GA. Unpublished results, 1982.
18. Karickhoff, S.W. Chemosphere, 1981, 10, 833.
19. O'Connor, D.J.; Connolly, J.P. Water Research, 1980, 14, 1517.
20. Voice, T.C.; Rice, C.P.; Weber, W.J., Jr. Environ. Sci. Technol., 1983, 17, 513.
21. Miller, C.T.; Weber, W.J., Jr. 186th National Meeting of the American Chemical Society, Washington, D.C., August, 1983; Abstr. PEST 73.
22. Hassett, J.J.; Means, J.C.; Banwort, W.L.; Wood, S.G. "Sorption Properties of Sediments and Energy Related Pollutants"; U.S. Environmental Protection Agency, Athens, Georgia. EPA-600/3-80-041, 1980.
23. Knott, G.; Reece, D. "MLAB", Division of Computer Research and Technology, National Institute of Health, Bethesda, MD 20014, 7th ed., 1977.
24. Wolfe, N.L. "Effects of Sediment Sorption on Abiotic Hydrolyses: III. Carboxylic Acid Esters", paper in preparation, 1984.
25. Stumm, W.; Morgan, J.J. "Aquatic Chemistry", 2nd Ed., John Wiley & Sons, New York, 1981; Ch. 10.

RECEIVED April 16, 1984

Investigation of Degradation Rates of Carbamate Pesticides

Exploring a New Detoxification Method

A. T. LEMLEY and W. Z. ZHONG

Department of Design and Environmental Analysis, Cornell University, Ithaca, NY 14853

G. E. JANAUER and R. ROSSI

Department of Chemistry, State University of New York at Binghamton, Binghamton, NY 13901

Base hydrolysis kinetic data are reported for ppb solutions of carbofuran, 3-OH carbofuran, methomyl and oxamyl. The results are compared with those reported previously for aldicarb, aldicarb sulfoxide, and aldicarb sulfone. Second order reaction rate constants, k_r, have been calculated and range from 169 liter min^{-1} $mole^{-1}$ for oxamyl to 1.15 liter min^{-1} $mole^{-1}$ for aldicarb. The order for rate of base hydrolysis is as follows: oxamyl >3-hydroxycarbofuran >aldicarb sulfone $\sim$ carbofuran >aldicarb sulfoxide > methomyl $\sim$ aldicarb. The activation energy for the base hydrolysis of carbofuran was measured to be $15.1 + 0.1$ kcal $mole^{-1}$, and is similar to the value previously reported for aldicarb sulfone. Rapid detoxification of aldicarb, a representative oxime carbamate pesticide, by in situ hydrolysis on reactive ion exchange beds is reported. Nucleophilic cleavage, acid catalyzed hydrolysis, and oxidation of aldicarb in dilute solution were achieved in batch and/or column experiments using macroporous reactive ion exchange resins. As in solution, nucleophilic cleavage proceeds faster than acid catalyzed hydrolysis. The basis for pursuing study of the latter mechanism is discussed.

Chemical degradation has been investigated by Shih and Dal Porto (1) and by Lande (2) under EPA auspices as an alternative approach (to landfill disposal) for the removal of pesticide residues. Among candidate reactions for the safe detoxification of pesticides, only alkaline hydrolysis was recommended. Several organophosphates and carbamates were identified as amenable to a degradation procedure using strong base/aqueous alcohol. The

0097-6156/84/0259-0245$06.00/0

main criterion for applicability was the virtual absence of toxic (degradation/reaction) products, which was met by 18 compounds including major pesticides such as malathion, phorate, and aldicarb. Although degradation/reaction products and their toxicological properties were identified wherever possible, there were many cases where such information was not available and chemical disposal was not recommended. The potential utility of alkaline hydrolysis was established, other possibilities such as acid hydrolysis explored, and additional research suggested by these authors (1,2).

Janauer et al. (3) showed recently that reactive ion exchange (RIEX) methods, previously employed for the preconcentration/isolation of inorganic trace species, can effectively degrade in situ (on suitable resins) representative organophosphates which are subject to nucleophilic attack. Thus, rapid decomposition/removal of organophosphorus pesticides can be accomplished by passing their solutions through a strong base resin bed in the free OH⁻ form. In one series of experiments, solutions containing 1-25 ppm of the pesticides malathion, guthion, diazinon, fenitrothion, and parathion were passed through short columns (ID = 0.70 cm) packed with strong base anion exchange resins (OH⁻ form) at flow rates varying from 2-5 ml/min which produced effluents containing less than 0.1% of the starting material (4). Additional column experiments with paraoxon essentially duplicated earlier results (5) so that the in situ degradation of organophosphates on reactive resins was shown to work with at least five different insecticides. Janauer also predicted that RIEX would be effective in detoxifying solutions containing carbamate pesticides. The discovery of the contamination of drinking water on eastern Long Island by the pesticide Temik, whose active ingredient is aldicarb [2-methyl-2-(methylthio)propionaldehyde-0-(methylcarbamoyl)-oxime], provided a case study for the application of strong base hydrolysis via RIEX to pesticide contaminated drinking water. The potential use of substitute crop protectants in areas where contamination has been discovered and the general lack of information with respect to the leaching properties and groundwater persistence of many carbamate and organophosphorus pesticides make it imperative to investigate the degradation behavior of a broad range of these compounds.

The collaboration of two research groups on this drinking water project was brought about through the intervention of Cornell's Center for Environmental Research. The initial seed money provided helped to lay the groundwork for the more ambitious project now being supported by the United States Environmental Protection Agency. Major research objectives of the project included:
   1. Development of improved methodology for
       separation and quantitation of aldicarb and
       similar carbamate pesticides.

2. Investigation of the kinetics of hydrolysis of aldicarb and other carbamate pesticides under a variety of conditions in solution.
3. Demonstration of the feasibility and efficiency of reactive ion exchange procedures for the detoxification of carbamates.
4. Development and field testing of an experimental Detoxification/Filter Unit by adapting a suitable conventional ion exchange device.

The first objective has been accomplished by the development of an HPLC procedure as reported by Spalik et al. (5) and GC/NPD procedures developed by Lemley and Zhong (6). The second and third objectives are being accomplished by fundamental solution studies and reactive ion exchange experiments conducted in parallel. Lemley and Zhong (7) determined recently the solution kinetics data for base hydrolysis of aldicarb and its oxidative metabolites at ppm concentrations and for acid hydrolysis of aldicarb sulfone. They have since (6) reported similar results for ppb solutions of aldicarb and its metabolites. In addition, the effect on base hydrolysis of temperature and chlorination was studied and the effect of using actual well water as compared to distilled water was determined. Similar base hydrolysis data for carbofuran, methomyl and oxamyl will be presented in this work.

Preliminary results of reactive ion exchange batch and column work will also be reported here. Column studies necessarily take more time to do and must rely on the wide range of data which can be obtained in solution. Values of k_{obs} obtained in solution are necessary for correlation with and prediction of column conditions. The final objective of this research, the development and testing of a detoxification/filter unit, will be pursued in the near future as soon as column conditions are sufficiently correlated with solution and batch RIEX results so as to permit optimization.

Experimental

Materials. Aldicarb standards were obtained from the United States Environmental Protection Agency (USEPA), Quality Assurance Section and from Union Carbide Corporation. Crystalline samples of carbofuran and 3-hydroxycarbofuran were supplied by the Agricultural Chemical Group of FMC Corporation. Reference standards of methomyl (99% pure) and oxamyl (99% pure) were obtained from USEPA. HPLC grade methanol was purchased from Burdick and Jackson, Inc. Methylene chloride used for bulk extractions of the carbamate pesticides in solution was recovered, distilled and reused. Analytical reagent grade chemicals and solvents were used in all experiments. Doubly distilled deionized water was used for solution rate studies. Deionized distilled water (DDW) was used for dilutions in reactive ion exchange experiments.

C_{18} SEP-PAK cartridges (Waters Associates) used in these experiments were pretreated by passing through 5 ml of HPLC-grade methanol followed by 10 ml of DDW. Reactive ion exchange work was performed using macroporous AG MP-1 strong base anion exchanger, 100-200 mesh, converted to the OH^- or $S_2O_8^=$ form (Cl or $SO_4^=$ were used as controls for sorption by resin) and AG MP-50 strong acid cation exchanger, 100-200 mesh, in the free H^+ form (Na^+ was used as a control for resin sorption) obtained from Bio Rad Laboratories. Solution blanks (no resin) were also used to determine the adsorption due to the reaction flask.

Degradation Rates. Procedures for determining the pseudo-first order, k_{obs}, and second order, k_r, rate constants for hydrolysis of carbamate pesticides in solution were similar to those reported for aldicarb/metabolites (7). A solution of known concentration of NaOH was added to a 200 ml flask and brought to the desired thermal equilibrium in a thermostated water bath. To this was added 1 ml of an appropriate concentration of pesticide solution such that the final concentration was known (25-200 ppb). The mixture was shaken immediately, and a slight excess amount of HCl was added at zero time and periodically thereafter (measured by stopwatch) to neutralize the base and stop the hydrolysis reaction. The solution was immediately transferred to a separatory funnel for extraction. The progress of base hydrolysis was followed directly by measuring the disappearance of the pesticide using the gas chromatographic procedures described in this paper. Procedures for ion exchange experiments in general and for determination of k_{obs}, the apparent in situ, pseudo-first order rate constant of hydrolysis for aldicarb on OH^- resin, in batch experiments has been previously reported (3).

Pseudo-first order rate constants, k_{obs}, were determined for the chemical degradation of dilute aldicarb solutions (200 ppb) by RIEX as follows: 1.0 gram of an air dried macroporous ion exchange resin, either AG MP-1 Cl^- form or AG MP-50 H^+ form, was weighed and allowed to swell in DDW for a minimum of four hours. The resin slurry was transferred to a fritted glass column (diameter = 0.70 cm.). Conversion to the appropriate counterion resinates was effected by passing 100 ml of a 0.1 N solution of either NaOH, Ba(OH)$_2$, $K_2S_2O_8$, $NaSO_4$, or $NaNO_3$ through the packed, air-bubble free resin bed which was then washed with 250 ml DDW. After air drying in the column, the resin was transferred to a 250 ml erlenmeyer flask to be used as the "reaction flask" and allowed to reswell in 50.5 ml of DDW.

The reaction flask was placed on a conventional waterbath shaker (Eberbach Corp., 180 oscillations/min.) and 50.00 ml of a 400 ppb aldicarb solution was added, and agitation started at 180 (linear) strokes per minute. The reaction was allowed to proceed for a specified period of time followed by filtration of the mixture and collection of the supernatant. The supernatant was then pumped through a C_{18} SEP-PAK and eluted with 2.00 ml of HPLC-grade methanol containing a small amount of acetic acid or acetate buffer (5). This eluate was passed through a Millipore clarification organic filter in a Luer-Lok syringe before analysis by HPLC (see below).

Analytical Procedures. The extraction procedure used for solution studies was based on the method supplied by Union Carbide Corporation (8), and was reported in previous work (6). Samples were analyzed using a Microtek GC with a Tracor NP detector as described by Lemley and Zhong (6). Conditions used for the various pesticides analyzed and the retention times obtained are detailed in Table I. In all cases a 4 foot x 4 mm I.D. glass column packed with 1.5% SP-2250/1.95% SP-2401 on 100/120 mesh Supelcoport was used. A glass injector was used for the analysis of carbofuran and 3-hydroxycarbofuran. Procedures were varied slightly for analysis of methomyl. The first three inches of the column were packed with 1.5% OV-17 on 100/120 mesh Supelcoport, and a glass injector was used without glass wool at the end.

Analyses for aldicarb (disappearance) subsequent to reactive ion exchange followed the procedures outlined by Spalik et al. (5). Analyses were performed using a Waters Model 6000A pump, a Model 440 absorbance detector fixed at 254 nm., a μ-Bondapack C_{18} column and a Model SRG Sargent recorder. In most experiments the eluting solvent was methanol:DDW (35:65) with 1% v/v acetic acid added to the solvent mixture. The flow-rate in HPLC runs was 1.0 ml/min. Peak heights were used to quantify aldicarb in some of the earlier experiments, but a Hewlett-Packard integrator was used later. A polystaltic pump (Buchler Instruments) was used to pump solution at a rate of 10 ml/min through a C_{18} SEP-PAK to preconcentrate it.

Results and Discussion

Pseudo-first order rate constants, k_{obs}, for the disappearance of pesticide in aqueous solution (doubly distilled deionized water) as a function of NaOH concentration were measured for solutions of carbofuran (30 ppb), 3-hydroxycarbofuran (200 ppb), methomyl (25 ppb), and oxamyl (25 ppb). These results are reported in Tables II-V. In each case excepting carbofuran a straight line with high correlation coefficient was obtained, confirming pseudo-first order behavior. There is a short plateau before the straight line portion in the carbofuran plots as shown in Figure 1. When experiments are performed at higher temperature, the plots become straight lines, indicating the presence of a short-lived intermediate. The k_{obs} values obtained from regression analyses of the slopes were plotted vs. hydroxyl ion concentration. These plots yielded straight lines passing through the origin for each species, and the second order

Table I. Chromatographic Conditions for Analysis of Carbamate Pesticides Extracted from Solution

	Gas Flow Rate (ml/min)				Temperature (°C)		Retention Time (min)
	He	H_2	Air	Injector	Column	Detector	
Aldicarb Sulfone	38	2.8	120	300	160	250	1.6
Carbofuran	35	3.6	120	220	190	250	4.9
3-OH Carbofuran	55	3.6	120	220	210	250	3.5
Methomyl	80	3.6	120	220	125	250	1.1
Oxamyl	35	2.8	120	275	130	250	1.9

Table II. Base Hydrolysis Rate Constants of Carbofuran (30 ppb) at 15°C

Concentration of NaOH x 10^3 (mole liter $^{-1}$)	k_{obs} x 10 (min $^{-1}$)	Standard Error (±)	r^2 (%)	k_r (liter min^{-1} mole^{-1})
3.73	1.16	0.001	99.8	31.1
5.59	1.67	0.002	99.8	29.9
6.51	1.95	0.001	99.9	30.0
8.14	2.58	0.001	100	31.7
9.77	2.95	0.005	99.5	30.2
			99.1	3.06 x 10 ± 0.6

Table III. Base Hydrolysis Rate Constants of 3-Hydroxycarbofuran (200 ppb) at 15°C

Concentration of NaOH x 10^4 (mole liter $^{-1}$)	k_{obs} x 10 (min^{-1})	Standard Error (±)	r^2 (%)	k_r x 10^{-2} (liter min^{-1} mole^{-1})
3.02	0.47	0.002	95.8	1.57
4.69	0.56	0.002	95.5	1.20
6.70	1.02	0.002	99.1	1.52
10.7	1.50	0.003	98.7	1.40
15.4	1.78	0.004	99.1	1.16
			96.0	1.19 x 10^2 ± 5

Table IV. Base Hydrolysis Rate Constants of Methomyl (25 ppb) at 15°C

Concentration of NaOH x 10^2 (mole liter^{-1})	k_{obs} x 10 (min^{-1})	Standard Error ($\pm$)	r^2 (%)	k_r (liter min^{-1} mole^{-1})
4.06	0.55	0.001	99.4	1.38
5.41	0.76	0.001	99.6	1.40
8.12	1.05	0.001	99.8	1.29
9.47	1.22	0.001	99.8	1.29
12.2	1.64	0.002	99.6	1.35
			99.2	1.32 $\pm$ 0.06

Table V. Base Hyrolysis Rate Constants of Oxamyl (25 ppb) at 15°C

Concentration of NaOH x 10^4 (mole liter^{-1})	k_{obs} x 10 (min^{-1})	Standard Error ($\pm$)	r^2 (%)	k_r x 10^{-2} (liter min^{-1} mole^{-1})
3.77	0.64	0.001	99.2	1.70
6.28	1.05	0.001	99.9	1.67
7.54	1.32	0.001	99.9	1.75
8.79	1.55	0.002	99.7	1.76
12.6	2.13	0.001	99.9	1.69
			99.8	1.69 X 10^2 $\pm$ 2

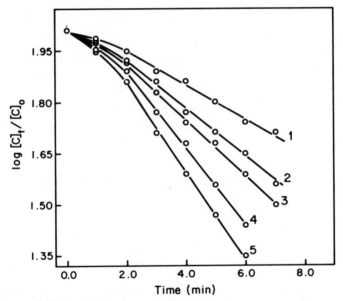

Figure 1.   Base hydrolysis of 30 ppb solution of carbofuran with various hydroxyl ion concentrations at 15 $^{\circ}$C.   1) 3.73 x 10^{-3}M, 2) 5.59 x 10^{-3}M, 3) 6.51 x 10^{-3}M, 4) 8.14 x 10^{-3}M, 5) 9.77 x 10^{-3}M.

reaction rate constant, k_r, was computed for each species as the slope of the respective line. A summary of the base hydrolysis second order reaction rate constants for all carbamate pesticides studied thus far is presented in Table VI.

The hydrolysis data collected for this group of carbamate pesticides have significance for several reasons. The k_{obs} values are expected to correlate directly with batch and column RIEX data such as those reported in this paper (vide infra). Since column and batch experiments with ion exchangers are more time consuming compared to solution studies, the data generated in solution become a useful base for decisions about future ion exchange experiments. In addition, the reaction rate constants calculated can be used to compare degradation of individual pesticides with each other. In cases where both the parent compound and oxidative metabolites have been studied e.g. aldicarb and carbofuran, the higher or highest oxidation species degrades more rapidly by hydroxide nucleophilic cleavage than does the parent compound. These results enable one to predict that a preoxidation treatment prior to hydrolysis should make a more efficient reactive ion exchange procedure.

A ranking of these pesticides with respect to ease of detoxification by hydrolysis can thus be used as a basis for determining treatment of drinking water, and can also be used to predict the relative environmental fate parameters. Assuming similar dependence of k_{obs} on OH^- concentration for environmental pH values, the rankings obtained in this study can be applied to environmental conditions and can be useful for pesticide application decisions.

Table VI. Base Hydrolysis Rate Constants of
Carbamate Pesticides
and Metabolites at 15°C

Name of Compound	k_r (liter min^{-1} $mole^{-1}$)	Standard Error ($\pm$)
Aldicarb	1.15	0.02
Aldicarb Sulfoxide	11.4	0.2
Aldicarb Sulfone	33.0	0.7
Carbofuran	30.6	0.6
3-Hydroxycarbofuran	119	5
Methomyl	1.32	0.06
Oxamyl	169	2

In addition to the above solution studies, experiments designed to study the effect of temperature on the base hydrolysis of carbofuran were performed at five temperatures between 5 and 35°C. The results are reported in Table VII. The studies were identical to the rate studies described above, and values for k_{obs} and k_r were calculated for each temperature. The results are similar to those obtained for aldicarb sulfone (6), i.e. a fifteen fold increase of k_r was found over the range of temperatures studied. The activation energy, E_a, was calculated as usual (Arrhenius plot) and had a value of 15.1 (± 0.1) kcal/mole.

Table VII. Base Hydrolysis Rate Constants of
Carbofuran at Different Temperatures

Temperature ($°C$)	k_r (liter min^{-1} mole^{-1})	Standard Error (±)	r^2 (%)
5	11.4	0.2	99.3
10	18.0	0.3	99.4
15	30.6	0.6	99.1
25	67.0	0.4	99.9
35	163	1.0	99.9

First reactive ion exchange studies have been conducted in batch, on column, and with several resin forms. For example, the degradation of 200 ppb solutions of aldicarb by nucleophilic cleavage and acid catalyzed hydrolysis was followed over time and plots of those data are shown in Figure 2. The k_{obs} values calculated as the slope of the lines are 6.6 x 10^{-2} min^{-1} and 9.3 x 10^{-4} min^{-1} for base and acid hydrolysis, respectively. As expected, the base hydrolysis was faster than the acid hydrolysis. This result was predicted by solution studies reported for aldicarb sulfone (7) and from recent results for aldicarb and aldicarb sulfoxide.(9) A minicolumn breakthrough study was performed to evaluate the practicality of the base hydrolysis RIEX method in a filter unit. A minicolumn (3.0 x 0.70 cm) containing 1.0 gram of Bio Rad AG MP-1 strong base anion exchange resin (100–200 mesh) converted to the OH⁻ form was used. A 1.1 ppm solution of aldicarb was passed through at a flow rate of ∿ 1 ml/min. Twenty-five ml samples of influent and effluent were removed for analysis after each of the volumes indicated in Table VIII. The concentration of aldicarb in these samples (not corrected for adsorption by resin) are reported in Table VIII. As can be seen, there appeared to be a "breakthrough" at approximately 2000 ml. Experiments are currently underway with both higher and lower concentrations and with aldicarb sulfoxide and aldicarb sulfone mixtures similar to those found in actual well water.

Table VIII. Breakthrough Capacity

Volume (ml) of Passed Through Through Resin Bed	Concentration A Influent (ppm)	Concentration A Influent (ppm)
50	1.1	0
100	1.1	0
200	1.1	0
400	1.1	0.1
600	1.1	0.1
650	1.0	0.1
800	1.0	0.1
1910	1.0	0.4
2343	1.0	0.5

Despite the faster base hydrolysis rate, the acid catalyzed reaction on resins is worth pursuing as a potential detoxification method of choice for two reasons. The acid hydrolysis of carbamate is a truly catalytic process leaving the protons available on the column for continuous reaction, whereas base hydrolysis uses up hydroxide ions and would require column recharge from time to time. In addition, the strong acid cation exchanger is a stable resin which can be stored for extended periods of time without changes, whereas the strong base anion exchanger absorbs CO_2 from air.

In order to study further the favorable aspects of in situ acid catalyzed hydrolysis, experiments were performed at different temperatures so as to evaluate the dependence of rate on temperature. Solutions of aldicarb were passed through a jacketed column around which water at 30, 40, or 50°C was circulating. The ion exchange bed (5 cm x 0.70 cm) contained 2.0 g of Bio-Rad AG MP-50 strong acid cation exchange resin (H^+, 100-200 mesh), and the solution flow rate was approximately 1.0 ml/min. The percent of initial aldicarb remaining at the end of the column for each temperature decreased from 76% at 30°C to 56% at 40°C and 35% at 50°C. Future temperature studies will be done in order to evaluate the practicality of temperature control in a detoxification filter unit.

Another interesting aspect of the RIEX studies to this date is the potential for precolumn or solution pretreatment oxidation. For solutions contaminated with a mixture of aldicarb and its metabolites, preoxidation to aldicarb sulfone would allow more efficient base hydrolysis since the degradation rate is faster for this metabolite (Table VI). Thus oxidizing agents were tested both in solution and on the resin to determine their potential for oxidizing aldicarb. The strong base anion

exchanger was converted to the S_2O_8 (peroxydisulfate) form and a
batch degradation study was performed with a 200 ppb solution of
aldicarb. A plot of ln [A] vs. time is shown in Figure 3. The
value for k_{obs} calculated from a linear regression of the slope
of the line is 2.7 x 10^{-2} min^{-1}. The products of this reaction
are aldicarb sulfoxide and aldicarb sulfone (not quantified)
indicating that oxidation is the predominant reaction. The fact
that a straight line is obtained indicates that one of these two
oxidation processes is rate determining. When compared to in
situ RIEX, aldicarb oxidation by S_2O_8 in free solution (no resin
present) was much slower.

These results are significant for several reasons. First,
peroxydisulfate, although a nucleophile, when exchanged on an
anion exchange resin, does not appear to effect nucleophilic
cleavage as does the hydroxide ion, but is predominantly an
oxidizing agent with aldicarb. The rate determining oxidation
step has a pseudo-first order rate constant which is almost the
same as the k_{obs} for base hydrolysis. Since this oxidation is
fairly rapid with the resin, there suggests itself the
possibility of using a preoxidation column prior to a strong base
anion exchange column (in series) for even more rapid chemical
degradation. Other oxidizing agents were also explored for use
with carbamates. Both perborate and peracetate decomposed when
sorbed on the resin. Hypochlorite was a particularly efficient
oxidant for aldicarb in solution, but did not work well in batch
when present on the resin. When solutions of aldicarb (5.0 ppm)
were mixed with $Ca(OCl)_2$ at concentrations of 4.7 ppm or higher,
there was complete degradation of aldicarb to the sulfoxide and
the sulfone as shown in Figure 4. A 1.9 ppm $Ca(OCl)_2$ solution
oxidized 40% of the aldicarb. There is the potential then for
using a bulk phase chlorination pretreatment step — similar to
classical water chlorination — for water contaminated with a
mixture of aldicarb and its metabolites. This method may also be
effective with other carbamate pesticides, and future work will
explore that possibility.

Conclusion

The second order alkaline hydrolysis rate constants were
determined for carbofuran, 3-hydroxycarbofuran, methomyl, and
oxamyl, and the activation energy was calculated for carbofuran
from results at different temperatures. This information may be
important in predicting the environmental fate of these species
when correlated with pertinent field data. Thus, one may be able
to model environmentally favorable and unfavorable conditions for
application of these carbamate pesticides. Furthermore, the
degradation of aldicarb in aqueous solution by different reactive
ion exchange resinates loaded with nucleophilic, acidic and
oxidizing counterions was shown to be feasible in situ, i.e. by
simple contact during passage over small resin beds. Combined

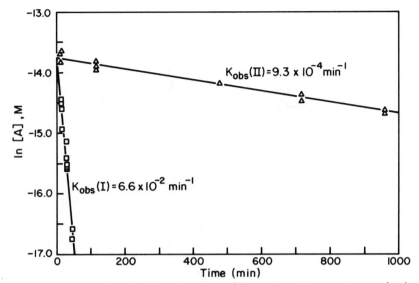

Figure 2. Nucleophilic cleavage (I) and acid catalyzed (II) hydrolysis of aldicarb by reactive ion exchange.

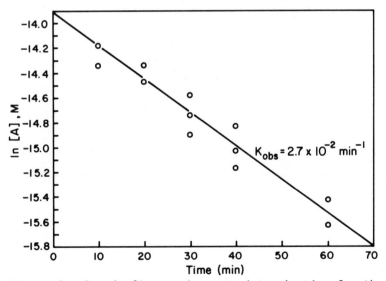

Figure 3. Pseudo-first order rate determination for the oxidative degradation of aldicarb by reactive ion exchange.

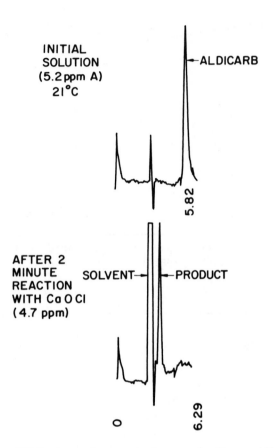

Figure 4. HPLC chromatogram before and after addition of
Ca(OCl)$_2$ to 5.2 ppm solution of aldicarb.

oxidative/hydrolytic systems and/or somewhat elevated temperature with acid catalyzed resinate systems may offer a simple means for detoxifying carbamate polluted drinking water. Correlation of fundamental solution parameters with such RIEX results will be important in the development of effective water detoxification systems based on RIEX. It is hoped that planned future work will eventually lead to adaptation of an optimized laboratory system to a preprototype field unit.

Acknowledgments

The authors gratefully acknowledge the financial support of the United States Environmental Protection Agency and the United States Department of Agriculture. They also appreciate the cooperation of the Union Carbide Corporation, FMC Corporation, and E.I. duPont deNemours & Company for providing samples.

Literature Cited

1. Shih, C.C.; Dal Porto, D.F. "Handbook for Pesticide Disposal by Common Chemical Methods," NTIS PB-252 864, Springfield, VA.
2. Lande, S.S; "Identification and Description of Chemical Deactivation/Detoxification Methods for the Safe Disposal of Selected Pesticides," NTIS PB-285, 208, Springfield, VA.
3. Janauer, G.E.; Costello, M.; Stude, H.; Chan, P; Zabarnick, S. in "Trace Substances in Environmental Health"; Hemphill, D.D., Ed.; University of Missouri: Columbia, 1980; Vol. XIV, pp.425-435.
4. Burrows, G.; Janauer, G.E., unpublished data.
5. Spalik, J.; Janauer, G.E.; Lau, M.; Lemley, A.T.; J. Chromatog. 1982, 253, 289-294.
6. Lemley, A.T.; Zhong, W.Z.; submitted for publication.
7. Lemley, A.T.; Zhong, W.Z.; J. Environ. Sci. Health, 1983, B18(2), 189-206.
8. "Determination of the Total Toxic Aldicarb Residue in Water", Union Carbide Corporation, 1980.
9. Lemley, A.T.; unpublished data.

RECEIVED March 5, 1984

Pesticide Availability
Influence of Sediment in a Simulated Aquatic Environment

ALLAN R. ISENSEE

Pesticide Degradation Laboratory, Beltsville Agricultural Research Center, U.S. Department of Agriculture, Beltsville, MD 20705

Sediment additions were made to model aquatic environments containing DDT equilibrated between sediment, water, and four species of aquatic organisms to determine if pesticide availability to biota was reduced through normal sedimentation events. Soil treated with ^{14}C-labeled DDT at 10, 100, and 1000 ppm was flooded and fish (Gambusia affinis), snails (Helosoma sp.), algae (Oedogonium cardiacum), and daphnids (Daphnia magna) were added. At two-week intervals, untreated sediment additions (equalling 1% of the water weight) were made. Samples of water and organisms, taken before and after sediment additions, were analyzed for DDT, DDE, and DDD and compared to equilibrated systems not treated with sediment. DDT content in water decreased 3-to 9-fold by the first sediment addition. Polar metabolites in water increased as DDT decreased. Fish were killed at the 1000 ppm level and daphnids succembed at 100 and 1000 ppm levels. Sediment additions substantially reduced the toxicity at lower treatment levels. Sediment additions decreased total ^{14}C by 6 to 13% in fish, 20 to 40% in algae, 45 to 50% in snails and 55% in daphnids (10 ppm rate). Measureable levels of DDT not diffuse through 1 cm or more of untreated soil into water in one year. Covering pesticide contaminated sediment with soil and sediment in situ is an effective contamination control method under certain aquatic conditions.

Pesticides that enter the aquatic environment from spills or improper treatment of manufacturing waste pose unusually difficult disposal problems. For example, the volume of contaminated and disposal methods are impractical and prohibitively expensive. In addition, the physical condition

of the site may severely limit the use of established techniques
and heavy equipment. Thus some in situ means of controlling the
contamination is highly desirable. One potential in situ
abatement procedure is to bury the contaminant in place under
soil or sediment. Highly insoluble, strongly adsorbed
contaminants would, ideally, be contained and biologically
isolated by this procedure. In addition, the anaerobic
conditions that develop under the overlaying soil or sediment
result in accelerated degradation of certain compounds.
Dechlorination of DDT under anaerobic conditions is well known
(1, 2). Dehalogenation of haloaromatic compounds has also been
demonstrated under anaerobic conditions (3).

The study was conducted to evaluate in situ sedimentation as
an abatement procedure. DDT was chosen as the test pesticide
since it is a persistent, water-insoluble compound that has
created difficult disposal problems and since a contaminated site
exists in Alabama where DDT residues estimated to exceed 500 tons
have been found in the bottom sediment of a 3-mile stream
segment.

Two experiments were conducted to (1) determine the effect
of soil (sediment) additions into a simulated aquatic environment
on the distribution and bio-availability of DDT and (2) determine
the extent of DDT diffusion through layers of untreated
sediment.

Methods and Materials

Microecosystem Chambers The aquatic microecosystem is shown in
Figure 1 and has been previously described (4). For this study,
160-g quantities of a sandy clay loam soil (58.4, 18.0, 26.6 and
1.94 % sand, silt, clay and organic carbon, respectively) were
treated with [^{14}C-ring] DDT (98+ % purity, 3.6 m Ci/mmole
specific activity) at 10, 100 and 1000 ppm and placed in the
bottom of 20-L capacity glass tanks (41 x 20 x 24 cm). The soil
was obtained from an uncontaminated location upstream from the
DDT contaminated site in Alabama. Three replicates of each rate
plus two control (160 g untreated soil) were prepared and then
flooded with 16-L activated carbon filtered tap water. One day
later 20 fish (Gambusia affinis), 20 snails (Helisoma sp.) and
1-g algae (Oedogonium cardiacum) were added to the tank and about
200 daphnids (Daphnia magna) were placed in the daphnid chamber
(1.5-L capacity tank with a stainless steel screen bottom to
restrict daphnid passage) which was suspended in the 20-L tank.
Water was continuously pumped (about 10 ml/min) into the daphnid
chamber which ensured uniform mixing of the water and transport
of food to the daphnids.

Untreated (no DDT) soil additions (160 g/tank) were made to
two of the three replicates at each concentration and to one of
the control tanks 14, 28 and 42 days after the addition of
organisms. Soil was first suspended in 2-L of water, then

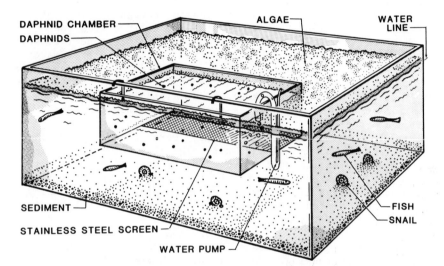

Figure 1. Microecosystem contains 16 L water and four species of organisms. The chamber measures 41 x 20 x 24 cm.

drained into tanks 3 cm below the water surface with the flow
directed horizontally to minimize disruption of the bottom
treated soil. A water volume of 16-L was maintained via a
properly located drain. The ecosystem was therefore designed to
simulate an inflow of water and sediment into a pesticide
equilibrated pond or lake.

Sampling and Analysis Water samples (triplicate 1-ml) were taken
at 2-day intervals and analyzed by standard liquid scintillation
(LS) methods for total ^{14}C. Larger water samples (100 ml)
were taken each week and extracted using "C 18 Sep-Paks" (Waters
Associates, Inc.)[1]. Samples were passed through Sep Paks at
2-4 ml/min. Recovery of ^{14}C-DDT plus metabolites from the
Sep Paks was achieved by extracting with 10 ml acetone followed
by 5 ml dichloromethane. DDT extraction efficiency from water
was 97+ %. Combined extracts (acetone plus dichloromethane) were
spotted on TLC plates (with unlabeled DDT, DDE and DDD) (20 x 20
cm GF-254, E. Merck, Darmstadt) and developed for 10 cm using
heptane:acetone (99:1). Plates were then scraped and total
^{14}C determined by LS. Tissue samples (two fish, two snails,
100 mg algae and about 50 mg daphnids) were taken 1, 3, 8, 15,
22, 29, 36, 43, 50 and 57 days after the start of the experiment.
Whole fish and snails were homogenized in acetonitrile, and the
homogenate was assayed directly by LS. Filtered samples of the
homogenates were analyzed by TLC as described above. Daphnids
were weighed, placed in LS vials, then ruptured with the cocktail
and analyzed directly by LS. Algae samples were oxidized to
determine total ^{14}C. DDT was acutely toxic to fish (1000 ppm
treatment) and daphnids (100 + 1000 ppm treatment) which
necessitated restocking. Only living fish and daphnids were
analyzed. Water and organism control samples were taken,
processed, and analyzed simultaneously with ^{14}C-DDT treated
samples. Final ^{14}C-DDT values were corrected using the
appropriate control. No ^{14}C in excess of background was
recovered from any control sample.

Diffusion Experiment Twenty gram quanities of the Alabama soil
were treated with [^{14}C-ring] DDT at 100 ppm and placed in the
bottom of 1 L beakers. Duplicate beakers containing the treated
soil were covered with 0, 1, 2, or 3 cm of untreated soil and
flooded with 800 ml water. Four cm of untreated soil in
duplicate beakers flooded with 800 ml water served as controls.
Triplicate 1-ml water samples were taken periodically. Two weeks
after flooding, 3 snails, 0.5 g algae and about 30 daphnids were

[1]Mention of a trade name or proprietary product does not
constitute a guarantee or warranty of product by the U.S. Dept.
Agric. and does not imply its approval to the exclusion of other
products that may be suitable.

added to each beaker. DDT was acutely toxic to the daphnids in
the 0 cm treatments which necessitated periodic restocking.

Results and Discussion

DDT in Water   The effect of sediment additions on DDT in water is
shown in Table I. There are two DDT data points for each
sampling day; the first is based on total ^{14}C analysis and
the second (in parentheses) is based on TLC analysis of the water
extracts. Total ^{14}C in water was not greatly affected by the
1% (160 g soil per 16-L water) sediment additions. Total ^{14}C
in the W/O tanks (not receiving sediment) increased continuously
with time while approximate plateau levels were maintained in the
W tanks (receiving sediment). In contrast, DDT levels (based on
TLC analysis) were greatly reduced. For example, on day 22, the
concentration of DDT in W tanks was 9, 5 and 3 times lower than
the concentration in W/O tanks for the 10, 100 and 1000 ppm
treatments, respectively. Total ^{14}C by comparison, was 1.1,
1.7, and 1.5 times lower for the same treatments. Clearly then,
input of sediment will reduce the concentration of DDT in
solution, particularly at the lower concentrations. However, the
results are complicated by the fact that DDT levels decreased
with time (after day 8 or 15) in all tanks, including those not
receiving sediment. TLC analysis of extracts from 100 ml water
samples partly explain this decrease (Figure 2). DDT
concentrations in the extracts averaged (for the three treatment
rates) 76, 67, 52 and 9% of the total recovered activity for days
3, 8, 15 and 22, respectively. For days 29 through 57, DDT
levels were below 5% of the recovered ^{14}C. Polar
metabolites, ^{14}C remaining at the TLC plate orgin, increased
as DDT decreased. These data suggest that the DDT in water was
degraded to polar metabolites and that little or no additional
DDT desorbed from the treated bottom sediments.

The slow increase in ^{14}C (in the W/O tanks) with time
may represent release of polar metabolites from the bottom
sediment. Total ^{14}C in water on day 57 represented 1 and 3%
of the total ^{14}C added to each tank at the start, for the W
and W/O sediment treatments, respectively. In addition, ^{14}C
recovered from the 100 ml water samples decreased from 88% of the
total ^{14}C (1 ml samples) on day 3 to 51% on day 57. These
results indicate that polar metabolites increased with time since
they are not recovered by Sep Paks. Only small amounts of DDE
and DDD (7 and 13% of total ^{14}C by day 8 and 15,
respectively) were detected in water.

The concentration of DDT in solution (Table I) often
exceeded the generally accepted water solubility of 2 ppb. Two
factors may account for these differences. First, the water
samples were not filtered before C 18 Sep Pak extraction.

Table I. Effect of Sediment Additions on DDT (ppb) in Water[a]

Ecosystem Treatment[b]

Days[c]	W Sed	10	W/O Sed	W Sed	100	W/O Sed	W Sed	1000	W/O Sed
3		0.5 (0.33)			6.8 (4.64)			69.3 (46.31)	
8		1.21 (0.71)			10.3 (5.54)			81.4 (48.65)	
15 S[d]		1.9 (0.71)			14.9 (7.07)			77.6 (42.17)	
22	1.8 (0.03)		2.1 (0.27)	10.9 (0.17)		18.0 (0.79)	59.4 (3.31)		89.9 (11.09)
29 S	1.7 (0.02)		2.1 (0.10)	12.2 (0.21)		19.9 (0.60)	63.8 (3.37)		86.8 (9.32)
36	1.3 (<0.01)		3.1 (0.13)	12.1 (0.06)		23.0 (0.07)	57.7 (1.21)		101.8 (5.78)
43 S	1.6 (<0.01)		3.3 (0.05)	11.5 (0.10)		26.2 (0.36)	72.8 (0.71)		138.6 (6.43)
50	1.3 (<0.01)		3.6 (0.05)	10.1 (0.13)		31.4 (0.33)	70.0 (0.68)		184.2 (8.49)
57	1.6 (<0.01)		3.7 (0.03)	9.9 (0.18)		31.2 (0.16)	70.6 (0.47)		234.2 (12.95)

a   Expressed as ppb DDT. Top figure based on total ^{14}C analysis, lower figure in (parentheses) based on TLC analysis of water extracts.

b   Concentration in ppm of DDT applied to 160 g of soil in each tank.

c   Days after flooding soil.

d   Sediment addition made immediately after sampling on days 15, 29, and 43.

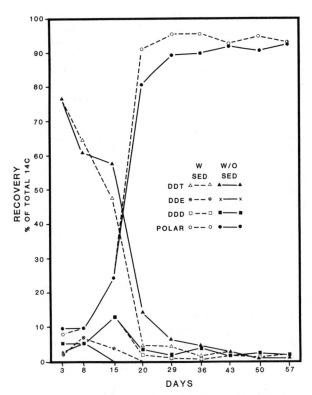

Figure 2. TLC analysis of water extracts showing relative distribution of DDT and metabolites with time.

The samples were visually clear but DDT could have been adsorbed
to very fine suspended sediment or colloidal material. Second,
the apparent water solubility may have been higher in the tanks
because of dissolved organic material, which has been shown to
bind hydrophobic organic compounds, such as DDT (5).

DDT in Aquatic Organisms   The concentration of DDT in the four
species of aquatic organisms (Tables II, III, IV and V) generally
reflected the treatment rates for the first 14 days, i.e. for
each 10-fold increase in treatment rate there was an approximate
10-fold increase in tissue concentration. This trend was not
followed for treatment rates that were acutely toxic (1000 ppm
rate for fish and 100 and 1000 ppm rate for daphnids).
Concentrations in the W sediment treatments decreased with each
sediment input relative to the W/O sediment treatment. By day
56, concentrations averaged 2 to 7, 2 to 4, 11 and 2 to 7 times
lower for the fish, snails, daphnids and algae, respectively.
Differences between the W sediment and W/O sediment treatments
were greater at the higher application rates. These results
indicate that all aquatic organisms responded to the decreasing
DDT levels in water.
        At the 1000 ppm rate, fish become lethargic within one day,
started to die by day 3 and were all dead by day 15. Additional
fish, added on day 16, continued to die, but at a much slower
rate in the W sediment tanks (Table II). A few fish also died in
the 100 ppm tanks, necessitating the addition of more fish on day
50. TLC analysis of the fish extracts show that DDT decreased
slowly with time, possibly accounting for the toxic response
(Figure 3). On day 3 about 90% of the total ^{14}C extracted
from fish was DDT which decreased slowly to 60 to 70% by day 57.
DDD and polar metabolite increased slowly with time. The
concentration of DDT in fish from the W sediment tanks remained
10 to 20% higher than fish from the W/O sediment tanks.
        Snails were not affected by DDT and tended to accumulate
less DDT than did fish for the same treatments (Table III).
Figure 4 partly explains these responses. TLC analysis indicated
that DDT was rapidly degraded by snails, decreasing from about
87% of the total ^{14}C on day 8 to 10% to 25% by day 57. Polar
metabolites and DDD increased with time.
        DDT was very toxic to daphnids. As a result, restocking was
required at the 100 ppm treatment on days 5, 17, 31 and 40 (Table
IV). Repeated restocking at the 1000 ppm treatment failed to
maintain a population for more than one day while daphnids
survived and reproduced at the 10 ppm rate. Daphnids, because of
their large surface-area-to-mass ratio, tended to respond to
changes in solution DDT levels more rapidly than did fish or
snails. Algae were, of course, not affected by DDT (Table V) and
the concentration of ^{14}C in the tissue responded to the
solution concentration in a very direct and rapid manner.

Table II. Effect of Sediment Additions on DDT Residues (ppm) in Fish (Gambusia affinis)[a]

Days[c]	Ecosystem Treatment[b]					
	10 W Sed	10 W/O Sed	100 W Sed	100 W/O Sed	1000 W Sed	1000 W/O Sed
1	0.3±0.1[d]		3.2±0.6		33.6±9.0	
3	1.0±0.2		10.6±2.2		112.3±23.0	
7	1.4±0.1		20.8±1.9		218.0±31.4	
14 S[e]	2.8±0.9		27.6±9.1		411.0±29.2[f]	
19	2.2±0.2	2.8±0.6	23.5±4.1	27.0±5.2	12.7±5.0	15.6±0.2
28 S	2.4±0.3	3.1±0.1	23.4±4.6	29.1±5.4	41.1±12.3	71.7±11.1
33	2.3±0.2	3.3±0.4	20.5±4.6	26.7±1.2	25.8±2.5	65.3±3.1
42 S	1.9±0.1	3.6±0.0	17.1±2.8	28.0±1.3	48.7±13.8	No Fish
47	2.2±0.3	4.7±1.0	17.8±2.5[f]	33.5±1.5[f]	48.4±7.7[f]	No Fish
50	1.7±0.0	4.8	16.2±2.9	27.5±11.0	9.2±1.2	66.8±4.7

a Expressed as ppm DDT based on total ^{14}C analysis.
b Concentration in ppm of DDT applied to 160 g of soil in each tank.
c Days after introducing the fish.
d Average ± standard deviation, based on 6 fish (days 1, 3, 7 and 14), 4 fish each for W Sed and 2 fish each W/O Sed treatments.
e Sediment additions made immediately after sampling on days 14, 28, and 42.
f Fish added on days 16 and 50 to both W and W/O Sed tanks.

Table III. Effect of Sediment Additions on DDT Residues (ppm) in Snails (Helosoma sp)[a]

| Days[c] | Ecosystem Treatment[b] | | | | | |
| | 10 | | 100 | | 1000 | |
	W Sed	W/O Sed	W Sed	W/O Sed	W Sed	W/O Sed
1	0.3±0.1[d]		3.5±1.9		19.7±8.0	
3	1.1±0.4		11.4±1.7		73.6±5.6	
7	1.4±0.1		20.8±1.9		218.0±31.4	
14 S[e]	2.1±0.6		21.2±7.6		101.9±26.0	
19	0.7±0.1	1.6±0.0	10.9±2.9	18.9±3.8	82.0±16.4	82.5±12.2
28 S	0.7±0.1	2.1±0.5	5.4±0.9	14.8±0.1	38.7±9.7	79.4±4.8
33	0.4±0.0	1.3±0.1	2.9±0.2	9.3±2.1	20.8±1.8	65.8
42 S	0.7±0.0	1.0±0.2	2.7±0.3	8.3±0.1	19.0±2.0	86.8±8.6
47	0.4±0.0	1.1±0.1	2.1±0.4	6.8±0.2	17.1±1.7	56.1±1.1
56	0.3±0.0	0.6±0.7	2.2±0.2	8.6±0.0	18.9±2.3	76.3±11.1

a Expressed as ppm based on total ^{14}C analysis.
b Concentration in ppm of DDT applied to 160 g of soil in each tank.
c Days after introducing the snails.
d Average ± standard deviation, based on 6 snails (days 1, 3, 7 and 14), 4 snails for W Sed and 2 snails for W/O Sed treatments.
e Sediment additions made immediately after sampling on days 14, 28, and 42.

Table IV. Effect of Sediment Additions on DDT Residues (ppm) in Daphnids (Daphnia magna)[a]

| Days[c] | Ecosystem Treatment[b] | | | |
| | 10 | | 100 | |
	W Sed	W/O Sed	W Sed	W/O Sed
1	2.3±0.1[d]	2.3	16.0±1.8	16.0[e]
3	4.2±0.3		27.6±3.7[e]	
7	7.1±1.6		19.4±3.0	
14 S[f]	4.9±0.6		Dead	
19	2.0±0.2	2.3	8.3±0.1	16.0[e]
28 S	N.S.[g]	N.S.[g]	N.S.[g]	
33	0.8±0.4	2.4	2.1±0.2	17.0[e]
42 S	0.5±0.2	1.1	1.4±0.1	1.2[e]
47	0.3±0.1	1.1	1.7±0.2	8.1
56	0.2±0.1	2.4	N.S.	N.S.

a Expressed as ppm based on total ^{14}C analysis.
b Concentration in ppm of DDT applied to 160 g of soil in each tank.
c Days after introducing daphnids.
d Average ± standard deviation, based on three-40 mg samples (days 1, 3, 7 and 14) and two-40 mg samples for the W Sed samples.
e Daphnids restocked on days 5, 17, 31 and 40.
f Sediment additions made immediately after sampling on days 14, 28 and 42.
g N.S. = not sampled.

Table V. Effect of Sediment Additions on DDT Residues (ppm) in Algae (Oedogonium cardiacum)[a]

| Days[c] | Ecosystem Treatment[b] | | | | | |
| | 10 | | 100 | | 1000 | |
	W Sed	W/O Sed	W Sed	W/O Sed	W Sed	W/O Sed
1	1.0±0.3[d]		17.0±8.7		139.4±26.0	
3	2.9±0.8		26.8±3.8		180.9±46.5	
7	1.9±0.3		22.7±4.7		280.2±87.1	
14 S[e]	3.9±1.0		26.0±3.9		291.3±62.0	
19	3.1±0.9	3.7±0.2	17.0±3.8	44.0±3.2	180.3±37.4	322.0±9.5
28 S	2.3±0.2	3.6±0.1	16.4±4.3	25.9±1.1	105.3±6.6	153.1±2.8
33	1.4±0.1	3.5±0.0	9.9±1.7	33.8±0.1	75.1±17.9	143.0±1.4
42 S	1.5±0.2	4.7±0.2	9.0±1.4	21.9±0.5	51.1±9.8	235.0±28.8
47	1.1±0.1	3.8±0.0	9.4±1.0	33.6±0.3	34.7±3.2	247.4±14.3
56	1.6±0.2	3.3±0.0	8.4±1.6	43.0±2.0	46.8±3.7	310.8±4.8

a Expressed as ppm based on total ^{14}C analysis.
b Concentration in ppm of DDT applied to 160 g of soil in each tank.
c Days after introducing algae.
d Average ± standard deviation, based on six-20 mg samples (days 1, 3, 7 and 14), four-20 mg samples for W Sed and two-20 mg samples for W/O Sed treatments.
e Sediment additions made immediately after sampling on days 14, 28, and 42.

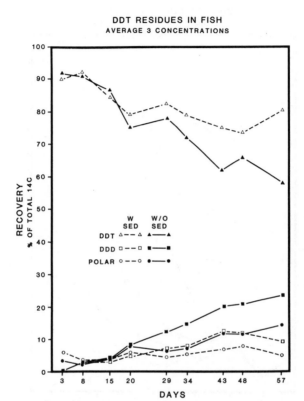

Figure 3. TLC analysis of fish extracts showing relative
distribution of DDT and metabolites with time.

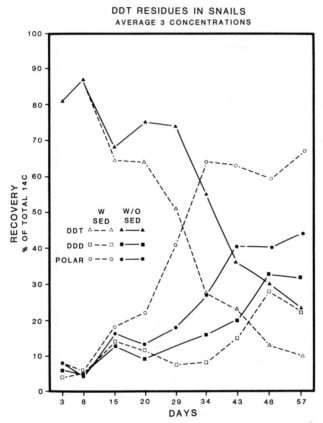

Figure 4.   TLC analysis of snail extracts showing relative
distribution of DDT and metabolites with time.

Diffusion Experiment   Results of the diffusion experiment are
shown in Table VI.   One or more cm of untreated soil covering 20
g of soil treated with 100 ppm ^{14}C-DDT was very effective in
preventing toxic concentrations of DDT from diffusing into water
for one year.   If any DDT did diffuse through the soil into
water, the concentration was not sufficiently high to affect the
survival or reproduction of daphnids.   A 60% reproductive
impairment has been reported when daphnids were exposed to 100
ng/L DDT (6).   Therefore, on the basis of the daphnid bioassay,
the concentration of DDT in water over the 1 cm of soil was at or
below 100 ng/L.   On the other hand, where untreated soil did not
cover the DDT layer, daphnids never survived more than 7 days.
This result is very similar to those from the microecosystem
experiment.   The 1-ml water samples indicated a total DDT
concentration of 10 to 20 ppb.   In addition, TLC analysis of
treated soil extracts after one year showed the expected
conversion of DDT to DDD, but only when covered by 1 or more cm
of soil.   For the uncovered soil, 87% of the radioactivity was
DDT.   Apparently, 1 cm of soil was sufficient to produce the
anaerobic conditions known to be necessary for conversion of DDT
to DDD (1, 2).
   This study shows that "sedimentation" may be a viable method
to isolate a pesticide contaminant under certain aquatic
situations.   If a persistent hydrophobic pesticide or other
organic contaminant is already present and adsorbed to stream or
lake bottom sediment, then additions of uncontaminated soil or
sediment will reduce the equilibrium concentration in water and
significantly reduce the exposure potential to aquatic organisms.
For the method to work, one assumes that the covering layer is
sufficiently thick to reduce exposure to burrowing benthonic
organisms and, for streams, that some means be employed to
protect the covering layer from erosion under high runoff
conditions.

Table VI. Movement and Degradation of DDT as Affected by
Sedimentation

% of Extracted ^{14}C[a]

treatment[b]	Daphnid bioassay[c]	Polar	DDD	DDT
Control	S and R	-	-	-
DDT W 0 cm Soil	D	1.1	5.7	86.8
DDT W 1 cm Soil	S and R	4.8	84.4	4.7
DDT W 2 cm Soil	S and R	5.3	80.5	6.5
DDT W 3 cm Soil	S and R	4.8	85.3	4.0

a. After 1 year treated soil Soxhlet extracted, then analyzed
by TLC.
b. Twenty g soil treated with 100 ppm ^{14}C-DDT, placed in 1 L
beaker covered with 0, 1, 2 or 3 cm untreated soil, then
flooded with 800 ml H_2O. Control = no DDT plus 4 cm soil
and H_2O c. Results of adding 20-30 daphnids to each
beaker. S and R= survived and reproduced; D = died within 5
days after each introduction for 1 year.

Literature Cited

1. Guenzi, W. D.; Beard, W. D. Soil Sci. Soc. Amer. Proc. 1968, 32, 522-524.
2. Parr, J. F.; Smith, S. Soil Sci. 1974, 118, 45-52.
3. Suflita, J. M.; Horowitz, A.; Shelton, D. R.; Tiedje, J. M. Science 1982, 218, 1115-1117.
4. Isensee, A. R. in "The Handbook of Environmental Chemistry, Vol. 2/Part A"; Hutzinger, O., Ed.;Springer-Verlog: New York, 1980 p. 231-245.
5. Carter, C. W.; Suffet, I. H. Environ. Sci. Technol. 1982, 16, 735-740.
6. Johnson, W. W.; Finley, M. T. in "Handbook of Acute Toxicity of Chemicals in Fish and Aquatic Invertebrates" U.S. Department of the Interior, Fish and Wildlife Service, Resource Pub. 137, Washington, D.C., 1980; p. 25.

RECEIVED March 9, 1984

Organophosphorus Pesticide Volatilization
Model Soil Pits and Evaporation Ponds

PAUL F. SANDERS and JAMES N. SEIBER

Department of Environmental Toxicology, University of California at Davis, Davis, CA 95616

A simple environmental chamber was used to measure volatilization of mevinphos, diazinon, methyl parathion, malathion, and parathion from model soil pit and evaporation pond disposal systems, and experimental results were compared to mathematical model predictions. Experimental volatilization rate constants were determined for the pesticides from water. Henry's law constants gave good estimates of their relative volatility, and absolute volatilization rates could be predicted from measured water loss rates or wind speed measurements. For pesticides with binding constants to soil greater than one, volatilization was much more rapid from the evaporation pond than from the soil pit. The EXAMS computer program gave good estimates of volatilization rates from water and water-soil systems. Triton X-100 decreased volatilization rates of low solubility pesticides from water.

It has been estimated that over 400,000 m^3 of dilute waste pesticide solution are generated in the United States annually ($\underline{1}$). While the bulk of these waste solutions are used legally as spray diluent, some are disposed of by chemical or biological treatment, incineration, or in soil pits and evaporation ponds ($\underline{1-3}$). Because of the large volume of water involved, incineration is not a preferred method. Adsorption of pesticides onto media such as activated charcoal, as well as biological and chemical treatment, are feasible methods, but they require frequent monitoring and maintenance. Evaporation ponds and soil pits have the advantages of less maintenance, applicability to a broad range of chemicals, and the ability to reduce the volume of waste via water evaporation. ($\underline{1-3}$). In addition, these latter two methods have been estimated to be the least expensive on a per gallon basis of waste ($\underline{1}$). This is of considerable importance because the wastes are

0097-6156/84/0259-0279$06.00/0

generally disposed of by the farmer or applicator on site by the
cheapest and simplest method available.

Despite the advantages, there is concern over the use of such
containment methods because the fate of pesticides put into such
sites is not well known (1). One such fate process is volatilization from the disposal site. Organophosphorus pesticide volatilization from water and soil is relatively uninvestigated, and if
this route of loss occurs to an appreciable extent from disposal
sites, a local respiratory hazard may exist.

In this paper, the volatilization of five organophosphorus
pesticides from model soil pits and evaporation ponds is measured
and predicted. A simple environmental chamber is used to obtain
volatilization measurements. The use of the two-film model for
predicting volatilization rates of organics from water is illustrated, and agreement between experimental and predicted rate constants is evaluated. Comparative volatilization studies are
described using model water, soil-water, and soil disposal systems, and the results are compared to predictions of EXAMS, a
popular computer code for predicting the fate of organics in
aquatic systems. Finally, the experimental effect of Triton X-
100, an emulsifier, on pesticide volatilization from water is
presented.

Model Pesticides

Five organophosphorus pesticides were chosen that could be iso-
thermally and simultaneously analyzed by gas chromatography using
an N-P TSD detector. They are all currently commercially used
and exhibit a wide range of physicochemical properties (Table I).
Also influencing the choice of these pesticides was the fact that
volatilization data measured from soil and water under controlled
laboratory conditions are scarce for methyl parathion, parathion,
and diazinon (14-17), and are not available for malathion and
mevinphos. Technical mevinphos (60% E-isomer, Shell Development
Co.), diazinon (87.2%, Ciba-Geigy Corp.), and malathion (93.3%,
American Cyanamid), and analytical grade methyl parathion (99%,
Monsanto) and parathion (98%, Stauffer Chemical Co.) were used.

Laboratory Model

Model Disposal System. The specific disposal systems modeled use
lined pits as described by others (1-3). The lining is usually
rubber or concrete, and is used to prevent pesticide solution from
leaching to the surrounding area. Because of the impervious
liner, the only transport route for parent pesticide is volatilization, providing the liner remains intact. The simplicity of
these systems allowed the use of a crystallizing dish as a model
disposal pit. The dish (50 x 100 mm; inside depth, 0.044 m;
inside diameter, 0.095 m; capacity, 310 ml) was filled to the brim
with water or soil containing the desired amount of pesticide.

Table I. Physicochemical Properties of
Model Pesticides at 22°C

Pesticide	Molecular Weight	Water Solubility[a]	Vapor Pressure[b]	Hydrolysis Rate Constant[c]
Mevinphos	224	miscible[d]	2200[f]	34.5[i]
Diazinon	304	68.8[d]	162[g]	2.58[j]
Me Parathion	263	37.7[e]	11.2[h]	2.97[k]
Malathion	330	143[d]	8.24[g]	250[l]
Parathion	291	12.4[e]	6.05[h]	5.08[j]

[a] ppm
[b] mm Hg x 10^6
[c] hr^{-1} x 10^4, pH 8.5
[d] Bowman et al. (4)
[e] Bowman et al. (5)
[f] Freed et al. (6)
[g] Estimated from Spencer (7)
[h] Calculated from Spencer et al. (8)
[i] Calculated from Worthing (9)
[j] Calculated from Harris (10) and Faust et al. (11)
[k] Calculated from Smith et al. (12)
[l] Calculated from Harris (10) and Wolfe et al. (13)

Environmental Chamber. The model disposal system was placed in an environmental chamber (Figure 1). Details of the chamber and its use are discussed elsewhere (18). The chamber was designed so that it was easy to switch model waste dumps by simply changing crystallizing dishes. It was constructed from two five-gallon Pyrex bottles with their bottoms cut off. The crystallizing dish rested in a Pyrex tray contained inside the chamber. An air dispersion tube supplied a variable laminar air flow across the surface of the dish. The air, containing volatilized pesticide (and water), exited from the chamber through XAD-4 resin, which trapped volatilized pesticide. To determine the amount volatilized, the resin was extracted with ethyl acetate, the solvent volume reduced, and the pesticides were analyzed by gas chromatography. Volatilization data reported from the environmental chamber were not corrected for volatilized pesticide recoveries. Recovery studies previously run (18) gave percent recoveries as follows: mevinphos, 72±6%, diazinon, 83±5%, methyl parathion, 77±5%, malathion, 76±8%, and parathion, 76±6%. Inlet air was humidified by passing it through a column which had water trickling down it over glass rings (18). By controlling the fraction of air that passed through the column, the relative humidity could be varied and controlled.

Although it could also be varied, the air flow rate through the chamber was always set at 20 lpm, which gave an air turnover time of 1.7 minutes. Wind speed measurements were taken using a

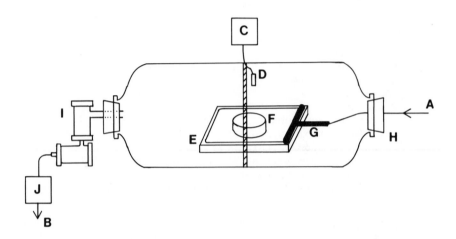

Figure 1. Environmental chamber. A, air entrance; B, air exit; C, hygrotherm; D, temperature and humidity probe; E, Pyrex tray; F, crystallizing dish; G, air dispersion tube; H, #12 rubber stoppers (covered with aluminum foil); I, pesticide vapor traps; J, flow meter. Reproduced with permission from Ref. 18. Copyright 1983, Pergamon Press.

Datametrics Model 100VT Air Flow Meter. Measurements were taken at various locations above the soil or water surface at a height of 0.8 cm, where the laminar air flow velocity was greatest. Depending on the probe location relative to the air dispersion tube, the measured wind speed varied from 0.5 to 1.5 m/s, with an average of 1 m/s. At greater heights above the surface, the air flow rate was much lower and the air flow patterns were unknown.

Two-Film Model for Volatilization of Organics from Water

Volatilization of a chemical from water is a first order process (19):

$$Rate = kC \qquad (1)$$

where k is the first order rate constant (time^{-1}), and C is the concentration of chemical in water. Models used to predict volatilization rates have been based on the two-film theory (19), the Knudsen equation (15), and on principles based on evaporation from surface deposits (2). The two-film theory is a simple and widely used kinetic model for calculation of volatilization rate constants of organic chemicals from water, originally applied to environmental volatilization by Liss and Slater (20) and later by Mackay and Leinonen (21) and Smith et al. (19). According to the two-film theory, volatilization from a water body depends on molecular diffusion of a chemical through thin films of air and water on either side of the interface. It is assumed that the chemical is at equilibrium across the interface as determined by the Henry's law constant for the chemical. Consideration of the above concepts leads to the following equation for the first order rate constant:

$$k = (1/L)((1/k_1) + (RT/H k_g))^{-1} \qquad (2)$$

where L is the solution depth (m), k_1 and k_g are the liquid and gas film mass transfer coefficients, respectively (m/s), R is the gas constant (atm m^3 K^{-1} mol^{-1}), T is the temperature (K), and H is the Henry's law constant (atm m^3 mol^{-1}). The Henry's law constant is defined as the ratio of the partial pressure of a solute above water to its solution concentration at equilibrium, and for low solubility compounds can be estimated by the ratio of its vapor pressure to its water solubility (22). Values for k_1 and k_g can be thought of as the velocities of chemical movement through the water and air films, respectively. They vary with environmental conditions, such as wind speed, water currents, and temperature, and with the diffusion coefficient of the chemical (23).

Henry's law constants for the pesticides used in this study were all less than 10^{-6} atm m^3 mol^{-1} (Table II). It has been shown (19) that in this case, the liquid film of Equation 2 is insignificant and

$$k = Hk_g/LRT \tag{3}$$

Furthermore, as long as environmental conditions remain constant, k_g will be nearly constant (diffusivity differences between pesticides will have only a minor effect) and variations in the rate constant should be largely controlled by the Henry's law constant (19).

A common method for estimating k_g values is to first estimate k_g for water, a well characterized reference compound, and then adjust this to the k_g of the compound of interest using the molecular weight adjustment of Liss and Slater (20):

$$k_{g_c} = k_{g_w} (18/M)^{\frac{1}{2}} \tag{4}$$

where M is the molecular weight of the pesticide, and c and w refer to the chemical of interest and water respectively. This adjustment corrects for diffusivity differences between compounds, which are inversely proportional to the square root of their molecular weights. The k_g value for water can be calculated using a measured water evaporation rate and an equation given by Smith (19):

$$k_{g_w} = NRT/(P°-P) \tag{5}$$

where $P°$ and P are the saturated and actual partial pressure (atm) of water vapor at temperature T (°K), R is the gas constant (atm m^3 mol^{-1} K^{-1}), and N is the water flux (mol m^{-2} hr^{-1}). Alternatively, k_g may be estimated by use of an empirical equation, such as that incorporated in the EXAMS computer code (24), that estimates k_g for water (m/hr) from the wind speed (m/s) at 10 cm height:

$$k_{g_w} = 0.1857 + 11.36 \text{ x WIND SPEED} \tag{6}$$

Wind speeds measured at other heights may be adjusted by assuming a logarithmic wind profile:

$$U_2 = U_1[\log(Z_2/Z_o)/\log(Z_1/Z_o)] \tag{7}$$

where U_2 and U_1 are wind speeds at heights Z_2 and Z_1, and Z_o is the roughness height. Although not illustrated here, pesticide k_1 and k_g values may be calculated directly from wind speed measurements and their respective diffusion coefficients (23).

In the present work, it was desired to 1) verify the prediction that the Henry's law constant controlled variations in the experimental volatilization rate constants under constant environmental conditions, and 2) compare experimental volatilization rate constants to predicted constants using the two methods for estimating k_g for water and the molecular weight adjustment procedure of Liss and Slater as discussed above.

Experimental and Predicted Volatilization Rate Constants

Pesticide solutions were prepared in tap water (pH 8.1±0.1) as reported previously (18). Initial pesticide concentrations were 180 ppm for mevinphos, 14 ppm for diazinon and methyl parathion, 24 ppm for malathion, and 3.5 ppm for parathion. Volatilization of the pesticides was measured each day for 7 days. The solution pH was 8.5±0.2, the relative humidity of the chamber air was controlled at 85±1%, and the chamber temperature average was 22± 2°C. Water (2 or 4 ml) was sampled, the resin in the trap was changed, and the water lost to evaporation (previously determined by weighing, 27±2 ml out of 310 ml) was replaced each day. Samples were prepared and analyzed as reported previously (18). Because of slow concentration decreases with time, low volatilization rates relative to hydrolysis rates in some cases, and small artificial losses of pesticide due to repeated water sampling, the most accurate method of determining volatilization rate constants was to divide the average pesticide concentration for that day into the average volatilization rate over the same period (Equation 1). Rate constants for the seven days were averaged. The entire experiment was performed in triplicate.

Although a logarithmic wind speed profile did not exist in our chamber, the measured wind speed was adjusted to an artificial wind speed of 2.21 m/s at 10 cm height using Equation 7.

Experimental and predicted volatilization rate constants for the five pesticides are listed in Table II. It should be noted that, despite low H values for the pesticides, experimental volatilization rates for diazinon and parathion are fairly rapid from water under the conditions of our tests ($t_{\frac{1}{2}}$ of 4.2 and 9.6 days, respectively). When compared to their hydrolysis rate constants (Table I), volatilization can be seen to be a more important route of loss than hydrolysis for diazinon, parathion, and methyl parathion. The relative volatilization rates reported here for diazinon and parathion are in good agreement with those reported by Lichtenstein (14).

It is apparent from Table II that variations in the experimental rate constants (k) are essentially controlled by the Henry's law constant, in agreement with the two-film theory prediction. A plot of k vs. H for the five pesticides gave an intercept of 5.4×10^{-4} hr^{-1}, a slope of 6.9×10^{3} mol/(hr atm m^3), and a correlation coefficient of 0.969. Thus, it seems that Henry's law values could be used to predict relative volatilization rates of the pesticides, and an absolute volatilization rate for one pesticide can be calculated if the volatilization rate is known for another and Henry's law constants are known for both:

$$rate_2 = rate_1 \times (H_2/H_1) \tag{8}$$

Except for mevinphos, agreement between experimental volatilization rate constants and rate constants predicted using k_g

Table II. Volatilization of Five
Organophosphates from Water

Pesticide	H^a	Experimental k^b ±1 Std. Dev.	Predicted k^b ±1 Std. Dev.[c] Method 1[e]	Predicted k^b_d ±1 Std. Dev.[d] Method 2[f]
Mevinphos	0.0518^g	0.36±0.06	0.076±0.006	0.04±0.02
Malathion	2.51	2.2±0.1	3.0±0.2	1.4±0.5
Me Parathion	10.3	13±1	14±1	6±3
Parathion	18.7	30±3	24±2	11±5
Diazinon	94.0	68±5	118±9	54±27

[a] $atm_1m^3 mol^{-1}$ x 10^8 (calculated from Table I)
[b] hr^{-1} x 10^4
[c] From uncertainty in water loss rate
[d] From uncertainty in measured wind speed
[e] Using measured water loss rate and Equation 5
[f] Using measured wind speed and Equation 6
[g] Vapor pressure was divided by density of
mevinphos (5.58 x 10^3 mol/m^3)

values from measured water loss rates is within a factor of two
(Table II). This close agreement indicates that volatilization of
low solubility organophosphates from water can be estimated from
water loss rates. The poor correlation for mevinphos is attributed to the method for estimating Henry's law constant, which is
only applicable to low solubility compounds. For compounds such
as mevinphos, a direct measurement of H is recommended. The gas
stripping apparatus of Mackay may be appropriate (25).

Except for mevinphos, agreement between experimental rate
constants and rate constants predicted using k_g for water, as
estimated from wind speed measurements, was within a factor of
three or better (Table II), although the laboratory measurements
were higher in all cases. At a given wind speed, it has been
observed that volatilization rates vary with the laboratory system
used, and are usually somewhat higher than those in the environment (23). Equation 6 was derived from wind tunnel experiments,
which mimic the real environment better than our chamber, so its
k_g predictions might be expected to be lower than those we
measured. The uncertainty in the wind speed is another important
consideration. As mentioned previously, the wind speed at 0.8 cm
height varied by ±50% depending on the probe location, and the
calculated wind speed at 10 cm height was an artificial value,
much higher than the actual wind speed at this height. Nonetheless, this approach was felt to be justified since air movement
immediately above the water surface would be expected to control
volatilization rates. Considering the above, the agreement
between experimental and predicted values is quite good. It appears

that methods to predict volatilization rates from wind speeds can
be used to estimate volatilization rates for a simple laboratory
system.   In this case, a method that required the assumption of a
logarithmic wind speed profile (Equation 7) was applicable to a
chamber that provided controlled air flow only immediately above
the water surface.

Volatilization from Water, Soil-Water, and Soil Systems

In order to ascertain the influence of soil on volatilization,
pesticides were incorporated into water, a water-soil mixture, wet
soil, and dry soil (1.7 (w/w) hygroscopic water still present),
and the percent volatilization that occurred in one day was measured.   For all four media, the initial pesticide amounts were held
nearly constant, and were approximately equal to the levels given
in the previous section.   The soil used was Reiff sandy loam (78%
sand, 16% silt, 6% clay, 2.8% organic matter, and bulk density
of 1.3 g/ml).   For soil samples, the desired amount of pesticide
dissolved in ethyl acetate (ca. 5 ml) was slowly dripped onto dry
soil with frequent shaking.   For wet soil, the soil was then
blended at low speed with a Waring blender with enough water to
saturate it (36% w/w) for one minute.   For the water and soil mixture, 73 g of dry soil were added to 286 ml of prepared pesticide
solution in the crystallizing dish.   The chamber conditions were
the same as in the previous section.   Soil samples (ca. 4 g) were
extracted for 1 minute with 250 ml ethyl acetate using a Waring
explosion proof blender, equipped with Polytron blades, at 80
volts.   The extracts were filtered through Whatman #1 filter paper
and concentrated for GLC analysis.   Results are uncorrected for
soil recoveries, which from triplicate determinations at 5 and 500
ppm spiking levels were found to be as follows: saturated soil,
87±4% (all pesticides); dry soil 76±4% (mevinphos), 91±6% (all
other pesticides).   Water samples from the water-soil system were
centrifuged to remove suspended material before extraction.
Simultaneous binding constants of the pesticides to Reiff sandy
loam at the concentrations used in this study were measured by
shaking 200 ml of prepared pesticide solution with 100 grams of
the soil for sufficient time to allow equilibration to occur (4
hours for mevinphos, 1 hour for the other pesticides).   The sand
was allowed to settle, and the silt and clay were removed from the
alkaline solution (pH 7.9) by centrifugation on a tabletop centrifuge.   The resultant decrease in the water concentration of the
pesticide was then measured by analyzing water samples taken
before and after mixing with soil.   Experimental binding constants
were as follows (ug/g soil)/(ug/ml water): mevinphos, 1.1±0.1;
diazinon, 4.3±0.4; methyl parathion, 4.0±0.3; malathion, 1.6±0.2;
parathion, 9±2.
    Diazinon, methyl parathion, and parathion, with binding
constants to soil considerably greater than 1, showed a decrease
in the percent pesticide volatilized in one day as the soil/water

ratio was increased (Figure 2). This was due to binding of the pesticides by the soil causing a decrease in the free pesticide concentration in water available for volatilization. Lichtenstein showed a similar decrease in volatilization for diazinon and parathion in the presence of soil (14). Mevinphos, with a binding constant of about 1, had a greater affinity for water than for soil (on a volume basis). As the soil/water ratio increased, it was crowded into a progressively smaller volume of water. Thus, its concentration in water and therefore its volatilization rate increased, although it was very low in all cases. Malathion, with a binding constant of slightly over 1, showed transitional behavior. Volatilization rates from dry soil were by far the lowest in all cases, because of the high binding capacity that dry soil has for pesticides in the absence of water (26).

The percent pesticide volatilized in one day from wet soil correlated positively with the factor [vapor pressure/(water solubility x binding constant)]. This factor has been reported to be linearly related to the volatilization rate of chemicals from soil surfaces (27). For pesticides with Henry's law constants and soil binding constants within the range studied, the factor is also approximately proportional to the fraction of chemical in soil air at equilibrium (28). In the present study, it was found that four of the pesticides had low factors, and less than 1% volatilized in 1 day (Table III). Diazinon, on the other hand, had a higher factor, and 2% of it volatilized. The use of this factor therefore does seem to have some merit for qualitative prediction.

Table III. Pesticide Volatilization from Wet Soil
Correlated with a Soil Volatilization Factor[a]

Pesticide	% Volatilized[b] ±1 Std. Dev.	Volatilization Factor[c] atm m^3 mol^{-1} x 10^8
Mevinphos	0.48±0.06	0.0471
Malathion	0.31±0.01	1.57
Parathion	0.8±0.2	2.08
Me Parathion	0.7±0.1	2.58
Diazinon	2.0±0.6	21.9

[a] Duplicate determinations
[b] In the first day
[c] Vapor pressure/(water solubility x binding constant)

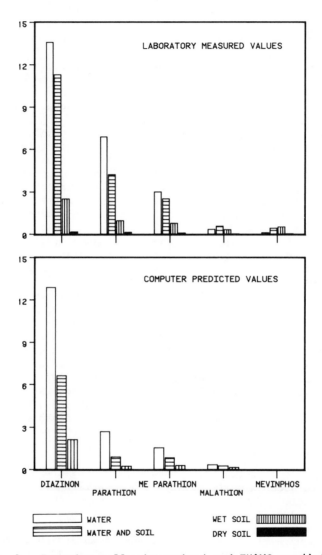

Figure 2.   Experimentally determined and EXAMS predicted
percents volatilized (in one day) for five organophosphorus
pesticides incorporated into water, water-soil, and soil
systems.   Computer predictions are not shown for mevinphos
or for dry soil.

Use of EXAMS with Model Disposal Systems

EXAMS (EXposure Analysis Modeling System) is an elaborate computer program that predicts the fate of organic chemicals in aquatic systems (24). Most input data can be easily measured, calculated, or obtained from literature sources. For this reason, the program is readily accessible to chemists for use as a predictive tool.

In the present study, EXAMS was used to calculate volatilization rate constants from water, wet soil, and a water-soil mixture. EXAMS uses the two-film theory to calculate volatilization rates from the 10 cm wind speed as discussed above. EXAMS requires as a minimum environment at least one littoral (water) and one benthic (sediment) compartment. A very small benthic compartment for the water system and a very small littoral compartment for the wet soil system (7.09 x 10^{-11} m^3 volume and 1 x 10^{-8} m depth in both cases) was used, so that these compartments and their input parameters had a negligible effect on the calculated rates. For the water-soil system, the same proportions were used as in the laboratory experiment. Transfer rates between soil and water were assumed to be rapid relative to volatilization rates, and were set as recommended in the EXAMS manual (24). The input data needed by EXAMS in order to calculate volatilization rates from a water-soil system, using parathion as an example, are shown in Table IV.

Percents volatilized in one day for the various media were calculated using initial pesticide amounts and the overall volatilization rate constants, obtained from the half life for volatilization as output by EXAMS. Mevinphos results are not included here, for as discussed previously, methods for calculation used in EXAMS are not appropriate for water miscible compounds. The experimental and computer predicted percents volatilized in one day are qualitatively similar (Figure 2). Quantitatively, experimental and predicted percents volatilized agreed within a factor of three for diazinon, methyl parathion, and malathion, and within a factor of five for parathion. Considering the fact that EXAMS was not intended for use with wet soil systems, these results are encouraging.

It should be noted that hydrolysis of these pesticides is expected to occur simultaneously with volatilization for the pesticides studied (Table I). Over a 7 day experiment, however, only malathion and mevinphos would be expected to hydrolyze to a significant extent. We determined the loss rate of mevinphos to be 0.0016±0.0002 hr^{-1} (t$_{\frac{1}{2}}$ = 18 days), and of malathion to be 0.011± 0.001 hr^{-1} (t$_{\frac{1}{2}}$ = 2.6 days) at 22±2°C, at pH 8.2±0.2 for a model evaporation pond by daily sampling of duplicate pesticide solutions (covered to prevent volatilization) for 7 days and plotting log concentration versus time. For both of these pesticides, then, degradation was a much more important route of pesticide loss from water than volatilization. The relatively slow loss rate of the other pesticides could not be determined in our 7 day

Table IV. EXAMS Volatilization Rate Constant Calculation
for a Water-Soil System: Input Data for Parathion

Parameter	Value
Molecular Weight (MWTG)	291.27
Water Solubility (SOLG(1))	12.4 ppm
Vapor Pressure (VAPRG)	6.05×10^{-6} mm Hg
Hydrolysis Rate Constants:	
Basic (KBHG(1,1))	99.4 hr^{-1} M^{-1}[a]
Neutral (KNHG(1,1), KNHG(2,1))	1.94×10^{-4} hr^{-1}[b]
Acidic (KAHG(1,1))	0
Binding Constant (KPSG)	9 (ug/g)/(ug/ml)
No. of Compartments (KOUNT)	2
Type of Compartment 1 (TYPEE(1))	L (Littoral)
Type of Compartment 2 (TYPEE(2))	B (Benthic)
Compartment Connection Variable (JTURBG(1))	1
Compartment Connection Variable (ITURBG(1)	2
Littoral Compartment Depth (DEPTHG(1))	0.034 m
Benthic Compartment Depth (DEPTHG(2))	0.0074 m
Littoral Compartment Volume (VOLG(1))	0.000254 m^3
Benthic Compartment Volume (VOLG(2))	0.000056 m^3
Surface Area (AREA(*), XSTURG(*))	0.00709 m^2
Bulk Density of Benthic Compartment (SDCHRG(2))	1.8 g/ml
Benthic Saturation Factor (PCTWAG(2))	136
Suspended Sediment (SDCHRG(1))	5 mg/l
Temperature (TCELG(*))	22°C
pH (PHG(*))	8.5
pOH (POHG(*))	5.5
Wind Speed at 10 cm Height (WINDG(1))	2.21 m/s[c]
Oxygen Exchange Constant (KO2G(1))	3.17 cm/hr[d]
Mixing Length (CHARLG(1))	0.021 m[d]
Dispersion Coefficient (DSPG(1))	1.5×10^{-4} m/hr[d]

[a] Free chemical only (29)
[b] Bound and free chemical (29)
[c] As calculated in earlier section
[d] As suggested in EXAMS manual (24)

experiment. However, literature hydrolysis rate constants for all
pesticides (from Table I) could be put into the EXAMS program.
The relative importance of the two processes in a model evaporation pond, along with the time for 97% loss of the applied pesticide (system purification time), were calculated (Table V). This
calculation confirmed that mevinphos and malathion dissipated
primarily by hydrolysis, with malathion the more rapid of these
two chemicals. For methyl and ethyl parathion, both processes
were significant, although volatilization was the dominant dissipation route. However, since both processes were relatively slow
for these pesticides, the purification time was fairly long.
Diazinon was predicted to be lost primarily via volatilization,
and the purification time was relatively short.

Table V. Relative Volatilization and Reaction
Rates as Predicted by EXAMS for a Model Evaporation Pond

Pesticide	% Hydrolysis	% Volatilization	Purification Time (days)
Malathion	99	0.6	6
Diazinon	4.6	95	26
Mevinphos	99.7	0.1	42
Parathion	31	68	89
Me Parathion	31	68	152

It is clear that computer codes such as EXAMS can be quite
useful for investigating the relative importance of fate processes
in evaporation ponds and soil pits.

Emulsifier Effect on Volatilization of Pesticides from Water

In order to more closely represent the volatilization environment
that would be encountered in an evaporation pond, Triton X-100, a
non-ionic emulsifier similar to those used in some pesticide formulations, was added to prepared pesticide solutions at 1000 ppm.
The presence of this emulsifier caused a decrease in the percent
pesticide volatilized in one day in all cases except for mevinphos
(Table VI). Three mechanisms are probably in operation here.
First, Triton X-100 micelles will exist in solution because its
concentration of 1000 ppm is well above its critical micelle concentration of 194 ppm (30). Pesticide may partition into these
micelles, reducing the free concentration in water available for
volatilization, which will in turn reduce the Henry's law constant
for the chemical (31). Second, the pesticides may exhibit an
affinity for the thin film of Triton that exists on the water surface. One can no longer assume that equilibrium exists across the
air-water interface, and a Triton X-100 surface film resistance

Table VI. Percent Volatilization of Five Organophosphorus
Pesticides in the Presence and Absence of Triton X-100

Pesticide	% Volatilized ±1 Std. Dev.	
	Triton Absent	Triton Present
Diazinon	12±1	5±1
Parathion	6±2	1.9±0.6
Me Parathion	2.6±0.2	1.3±0.1
Malathion	0.28±0.03	0.21±0.02
Mevinphos	0.12±0.02	0.13±0.02

Note:1000 ppm Triton X-100, triplicate 24 hour
experiments, chamber conditions same as
previously

must be added to the two-film equation (32). Third, the film of
Triton X-100 reduces the turbulence at the air-water interface,
thus reducing k_l and/or k_g values (31). Lichtenstein also noted
an inhibition of volatilization of diazinon, as well as several
other pesticides, when LAS detergent was added to buffered water
(14). However, he found that the detergent increased volatilization of parathion and dieldrin, indicating that surfactant effects
may be more complex than discussed here. It is clear that the
presence of formulation components in water complicates the estimation of pesticide volatilization from evaporation ponds.

Conclusions

A simple environmental chamber is quite useful for obtaining volatilization data for model soil and water disposal systems. It was
found that volatilization of low solubility pesticides occurred to
a greater extent from water than from soil, and could be a major
route of loss of some pesticides from evaporation ponds. Henry's
law constants in the range studied gave good estimations of
relative volatilization rates from water. Absolute volatilization
rates from water could be predicted from measured water loss rates
or from simple wind speed measurements. The EXAMS computer code
was able to estimate volatilization from water, water-soil, and
wet soil systems. Because of its ability to calculate volatilization from wind speed measurements, it has the potential of being
applied to full-scale evaporation ponds and soil pits.
    Whether one would want to maximize or minimize volatilization
rates from these systems is open to debate. The user may want to
maximize this rate, for it can help decontaminate the system and
keep pesticide concentrations under control. Alternatively, the
user or the public may want to minimize volatilization, to reduce
respiratory risks in the vicinity of the sites. The approach
taken here was simply to document what happens, so that the proper

choice of a disposal system can be made to fit the needs of a
specific user and site.

Acknowledgments

This work was supported in part by the National Institute of
Health training grant #PHS ES07059-06. Thanks go to EPA (Mr.
Larry Burns) for furnishing a copy of the EXAMS computer program.

Literature Cited

1.   SCS Engineers "Disposal of Dilute Pesticide Solutions";
     PB297-985, National Technical Information Service: Spring-
     field, VA, 1979; pp. 1-46.
2.   Hall, C.V.; Baker, J.; Dahm, P.; Freiburger, L.; Gordei, G.;
     Johnson, L.; Junk, G.; Williams, F. "Safe Disposal Methods
     for Agricultural Pesticide Wastes"; PB81-197584, National
     Technical Information Service: Springfield, VA, 1981.
3.   Egg, R.P.; Redell, D.L.; Avant, R. in "Treatment of Hazardous
     Waste: Proceedings of the 6th Annual Research Symposium";
     Shultz, D., Ed.; PB80-175094, National Technical Information
     Service: Springfield, VA, 1980; pp. 88-93.
4.   Bowman, B.T.; Sans, W.W. J. Environ. Sci. Health 1983, B18,
     221.
5.   Bowman, B.T.; Sans, W.W. J. Environ. Sci. Health 1979, B14,
     625.
6.   Freed, V.H.; Schmedding, D.; Kohnert, R.; Haque, R. Pestic.
     Biochem. Physiol. 1979, 10, 203.
7.   Spencer W. in "A Literature Survey of Benchmark Pesticides";
     George Washington University Medical Center, Science Commu-
     nications Division: Washington, D.C., 1976; pp. 95-96.
8.   Spencer, W.F.; Shoup, T.D.; Cliath, M.M.; Farmer, W.J.;
     Haque, R. J. Agric. Food Chem. 1979, 27, 273.
9.   Worthing, C.R., Ed. "Pesticide Manual"; British Crop Protec-
     tion Council: London, 1979; p. 363.
10. Harris, J.C. in "Handbook of Chemical Property Estimation
     Methods"; Lyman, W.J.; Reehl, W.F.; Rosenblatt, D.H., Eds.;
     McGraw-Hill: New York, 1982; pp. 7-42 and 7-43.
11. Faust, S.D.; Gomaa, H.M. Environ. Lett. 1972, 3, 171.
12. Smith, J.H.; Mabey, W.R.; Bohonos, N.; Holt, B.R.; Lee, S.S.;
     Chou, T.W.; Bomberger, D.C.; Mill, T. in "Environmental
     Pathways of Selected Chemicals in Freshwater Systems: Part
     II. Laboratory Studies"; EPA-600/7-78-074, U.S. Environmental
     Protection Agency, Environmental Research Laboratory: Athens,
     Georgia, 1978; pp. 271-5.
13. Wolfe, N.L.; Zepp, R.G.; Gordon, J.A.; Baughman, G.L.; Cline,
     D.M. Environ. Sci. Technol. 1977, 11, 88.
14. Lichtenstein, E.P.; Schulz, K.R. J. Agric. Food Chem. 1970,
     18, 814.

15. Chiou, C.T.; Freed, V.H.; Peters, L.J.; Kohnert, R.L. Environ. Int. 1980, 3, 231.
16. Metcalfe, C.D.; McLeese, D.W.; Zitko, V. Chemosphere 1980, 9, 151.
17. Smith, J.H.; Bomberger, D. AIChE Symposium Series 1979, 75, 375.
18. Sanders, P.F.; Seiber, J.N. Chemosphere, 1983, 12, 999.
19. Smith, J.H.; Bomberger, D.C.; Haynes, D.L. Chemosphere 1981, 10, 281.
20. Liss, P.S.; Slater, P.G. Nature, 1974, 247, 181.
21. Mackay, D.; Leinonen, P.J. Environ. Sci. Technol. 1975, 9, 1178.
22. Mackay, D.; Wolkoff, A.W. Environ. Sci. Technol. 1973, 7, 611.
23. Mackay, D.; Yeun, A.T.K. Environ. Sci. Technol. 1983, 17, 211.
24. Burns, L.A.; Cline, D.M.; Lassiter, R.R. "Exposure Analysis Modeling System: (EXAMS): User Manual and System Documentation"; U.S. Environmental Protection Agency, Environmental Research Laboratory: Athens, GA, 1981.
25. Mackay, D.; Shiu, W.Y.; Sutherland, R.P. Environ. Sci. Technol. 1979, 13, 333.
26. Spencer, W.F.; Farmer, W.J.; Cliath, M.M. Residue Rev. 1973, 49, 1.
27. Thomas, R.G. in "Handbook of Chemical Property Estimation Methods"; Lyman, W.J.; Reehl, W.F.; Rosenblatt, D.H., Eds.; McGraw-Hill: New York, 1982; pp. 16-25 to 16-27.
28. Bomberger, D.C.; Gwinn, J.L.; Mabey, W.R.; Tuse, D.; Chou, T.W. in "Fate of Chemicals in the Environment"; Swann, R.L.; Eschenroeder, A., Eds.; ACS SYMPOSIUM SERIES No. 225, American Chemical Society: Washington, D.C., 1983; pp. 197-214.
29. Wolfe, N.L. Pesticide Division Paper No. 73, 186th National American Chemical Society Meeting, Washington, D.C., September, 1983.
30. Stubicar, N.; Petres, J.J. Croat. Chem. Acta 1981, 54, 255.
31. Mackay, D.; Shiu, W.Y.; Bobra, A.; Billington, J.; Chau, E.; Yeun, A.; Ng, C.; Szeto, F. "Volatilization of Organic Pollutants from Water"; PB82-230939, National Technical Information Service: Springfield, VA, 1982; pp. 7-81.
32. Smith, J.H.; Bomberger, D.C.; Haynes, D.L. Environ. Sci. Technol. 1980, 14, 1332.

RECEIVED February 13, 1984

Potential Pesticide Contamination of Groundwater from Agricultural Uses

S. Z. COHEN and S. M. CREEGER—Office of Pesticide Programs, TS-769c, Environmental Protection Agency, Washington, DC 20460

R. F. CARSEL—Environmental Research Laboratories, Environmental Protection Agency, Athens, GA 30613

C. G. ENFIELD—Environmental Research Laboratories, Environmental Protection Agency, Ada, OK 74820

EPA began to emphasize work on ground water contamination by pesticides in 1979. Much monitoring has been done in this area since 1979, mostly by state agencies and, to a lesser extent, pesticide registrants, university scientists, and EPA. To date, as a result of agricultural use, a total of 12 different pesticides have been found in the ground water of 18 different states. The 12 chemicals represent seven different chemical classes. Despite significant limitations in the laboratory and field data, some generalizations about the key environmental fate parameters and field conditions are made which aid in predicting which compounds will leach to ground water.

Use of pesticides in the production of U.S. agricultural commodities is widespread; 370,455 metric tons of active ingredients, which corresponds to 70.3% of the total poundage of pesticide active ingredients used in the U.S., were applied to agricultural land in 1982 (1). With the worldwide increase in need for food and fiber (2), use of pesticides is expected to increase (3). Since nearly half of the U.S. population relies on ground water as their source of drinking water (4), potential for contamination of ground water due to pesticide use must be considered in the registration process.

In this paper, the environmental fate characteristics of those pesticides found to date in ground water as a result of agricultural use are summarized. Monitoring data are also summarized. From those summaries, and information on the sites of ground water contamination, it can be concluded which combination of pesticide chemical characteristics and use sites represent a high potential for ground water contamination. Other aspects of assessing potential for ground water contamination, such as study design and modeling, as well as interpretations of sorption, volatility, and photolysis data are also discussed.

The EPA began to emphasize work on ground water contamination by pesticides in 1979. The impetus for this effort was the finding of 1,2-dibromo-3-chloropropane (DBCP) and aldicarb (chiefly as the sulfoxide and sulfone metabolites) in ground water in various states (5-9). Prior to 1979, water monitoring studies had generally not focused on pesticides in ground water per se. The reasons are that: (1) monitoring had generally focused on urban rather than on rural, agricultural areas; (2) analyses were often for volatile/purgeable organics, and many pesticides are of low volatility and (3) reports of positive findings of organics in drinking water have not always distinguished between surface water and ground water systems. Another concern has been an increase in use of pesticides with higher water solubilities in recent years.

Overall, there have been many findings of pesticides in ground water. This paper presents the results of work done in this area over the last few years by state governments, pesticide registrants, university scientists, and the EPA. The results discussed here generally pertain to normal agricultural use, as opposed to waste dump sites, although some comments are made in the section on Chemical Characteristics, as well as in the Conclusions, on pesticide leaching from disposal sites.

Chemical Characteristics & Monitoring Data

In this section, two types of data are briefly summarized: 1) environmental transport and persistence, and (2) monitoring. The discussion is chemical specific. Some interesting concepts relevant to some of the chemical characteristics are developed later in the Discussion section. Results from this section are also used in the Conclusion section to derive some generalizations about pesticides leaching to ground water.

All data presented in this section were either generated under ambient conditions (pH 6 to 8, 15°-30°C) or extrapolated to ambient conditions, unless noted otherwise. In some cases, data was available only under conditions unlikely to be found in the environment. Definitions of terms are as follows: K_d = soil/water distribution coefficient; K_{oc} = K_d divided by the soil organic carbon fraction; mobility classes, given for three pesticides, are taken from Helling's work (10) which evaluates the movement of 40 pesticides in Hagerstown silty clay loam soil thin-layer plates; Henry's law constant (H) is a measure of the escaping tendency of dilute solutes from water and is approximated by the ratio of the vapor pressure to the water solubility at the same temperature (11). The significance of H values to environmental situations is presented in the Discussion section. Whenever possible, field dissipation half-lives are given, which include losses due to hydrolysis, microbial activity, volatilization, etc., in a lumped, pseudo-first-order process. It is realized that this is often an oversimplification, since many mechanisms for loss of pesticides in the environment are not first-order. The environmental fate summaries presented should only be considered as capsule summaries rather than extensive critical reviews. The well contamination incidents discussed are considered to be due to normal agricultural use of the pesticide and generalizations presented on pesticide leaching based on agriculturally related K_d values, rates of degradation, etc., are not always applicable to waste disposal sites. For example, it has been demonstrated that adsorption is less than anticipated, resulting in increased leaching, when pesticides are present at very high soil concentrations such as might exist at waste dump sites ([12,13]). Thus, the adsorption isotherms are not linear. This occurs even though the pesticide adsorption sites are not saturated as manifested by the fact that the data fit the Freundlich equation up to the solubility limits of the pesticides studied ([13]). There is also evidence that microbial degradation of high soil concentrations of pesticides is slower than lower soil concentrations ([14,15]). The impact of organic solvents on K_d values is less clear. Recent work ([16]) has demonstrated that organic solvents such as methanol, when present at high concentrations in hazardous waste sites, can significantly increase the leaching potential of organic solutes. However, it is not known whether organic solvents are usually present at high concentrations in landfill leachates. For example, recent analyses of leachates from above and below the liners of 11 landfills indicate that organic solvents were present at only ppm levels ([17]).

It is important to note that much of the monitoring data presented below results from studies which were not designed for statistical treatment of the results. Therefore, it is often difficult or nearly impossible to draw statistically reliable conclusions about the results from these studies. Also, key information such as depth to the water table and well construction are sometimes unavailable.

Soil core data are available for only six of the pesticides discussed in this paper. The six pesticides are: aldicarb; atrazine; 1,2-dibromo-3-chloropropane (DBCP); 1,2-dichloropropane (DCP); 1,2-dibromoethane (EDB); and simazine. Cores were always sampled at depths greater than one meter and the soil was characterized physically and chemically. The importance of soil core sampling in pesticide leaching assessments is presented in the Discussion section.

Alachlor. Alachlor is an acetanilide herbicide with a water solubility of 242 ppm ([18]) and a 2.2×10^{-5} torr vapor pressure ([18]). The approximated Henry's law constant (H) is 3.2×10^{-8} atm-m^3/mol. K_d values range from 0.6-8.1 in various types of soil ([19]). However, most K_d values are less than 4; the 8.1 value came from a measurement with a soil containing 11.7% organic matter. Alachlor's K_{oc} is reported as 120+49 ([19]) and 190 (no standard deviation given) ([20]).

Degradation of alachlor in various soils has been reported as follows: lab half-lives of 4-21 days (19), time for 90% dissipation in the field of 40-70 days (21), and half-lives in the field of 18 days (22) and 15 days (19). In addition, there is a report of the complete degradation of alachlor after 35 days in a model ecosystem (23).

Alachlor was found in two wells out of 14 sampled in Nebraska at about 0.04 ppb in a monitoring study which focused on atrazine high use areas (24). These corn growing areas were known to have sandy soils and a shallow, unconfined aquifer. The wells sampled were assumed to be representative of over 1,000 wells located in the study area.

Aldicarb and Products. Aldicarb is an oxime carbamate insect-icide/nematicide, with a water solubility of 6,000 ppm (25,26) and a vapor pressure of 7.95×10^{-5} torr (27). Its H value is 2.5×10^{-6} atm-m^3/mol. Its sulfoxide and sulfone products have higher water solubilities of 43,000 ppm and 7,800 ppm, respectively (26).

Aldicarb sulfoxide K_d values in a clay and in a silt loam, both containing 1.4% organic matter, were 3.3 and 0.34, respectively (28). K_d values for parent aldicarb would not be germane since it degrades so rapidly in soil under aerobic conditions to the sulfoxide and sulfone, but it would also have a K_d of less than four (19).

Aldicarb is stable to hydrolysis at pH 7.5 or lower and temperatures of 15°C or lower. The half-life at pH 7.5 and 15°C is 5 years and the half-life is 12.5 years at pH 5.5 and 5°C (typical ground water temperatures in the northern half of the U.S. at depths of 10-20 m, are 5-15°C). At pH 8.5 and 5°C, the half-life is about 3.8 years. At pH 7.5 or less, and temperatures between 5°C and 15°C, the sulfoxide has half-lives of 1-2.2 years and the sulfone has half-lives of 0.3-2.5 years; degradation is more rapid with increasing pH (29). Aldicarb sulfoxide and aldicarb sulfone hydrolysis products are nontoxic compounds (30-32).

The soil half-life derived in the lab for conversion of parent aldicarb to the sulfoxide and the sulfone is less than a week and degradation is more rapid with increasing soil moisture and temperature (27). However, volatilization may contribute to some of the loss of the parent (27). In the field, low levels of total residues have persisted for 6-12 months (19). This may be due to the aldicarb residues leaching beyond the zone of active microbial degradation. In addition, the aldicarb sulfoxide and sulfone degradation products of aldicarb might be reduced under anaerobic soil conditions to parent aldicarb or aldicarb sulfoxide, respectively, effectively resulting in persistence of aldicarb residues. This has not been demonstrated to occur with these degradation products but reduction under both aerobic and anaerobic conditions to the sulfide has been shown to occur with compounds containing the sulfoxide or sulfone moiety (33-36).

Based on analyses of well water from Suffolk County, New York, which show aldicarb sulfoxide and sulfone, (and generally 0 or <10% parent aldicarb) to be the only compounds in well water, and in an approximate 50:50 ratio, Union Carbide decided to analyze all ground water samples using a total residue method, i.e., by oxidizing all residues present to the sulfone.

Aldicarb residues, typically 1-50 ppb, have been found in ground water in New York, Wisconsin, Missouri, Florida, Virginia, Maine, Arizona, California, North Carolina, New Jersey, Oregon, Washington, and Texas (8,9,19,37). In New York, the highest concentrations in the water and the highest frequency of positives (94.4%) have been found within 300 meters of the application sites (8). This generalization is probably appropriate for the other states as well. In fact, most sampling has been done in high use areas. Some very limited well sampling in Oregon, Washington, Texas, and North Carolina has shown aldicarb residues at levels of 1-4 ppb, some well sampling around lily bulb farms in northern California has shown contamination up to 24 ppb, and sampling of 9 wells in south New Jersey has shown detectable residues in 3 of those wells at concentrations of 3, 4, and 50 ppb (19). As of 1982, 27% of 8,404 wells sampled in Long Island, New York contained detectable (>1 ppb) amounts of total aldicarb residues (8).

A study (38) conducted in a California orange grove showed subsurface soil water to contain the following total aldicarb residues (parent + sulfoxide + sulfone) at four hours after cessation of 48 hours of continuous irrigation:

Table I. Total Aldicarb Residues in Subsurface Soil Water of a California Orange Grove af Four Hours after Cessation of 48 Hours of Continuous Irrigation (38)

Rate (kg ai/ha)	Irrigation type	Soil depth (cm)	Residues (ppm)
11.2	furrow	30.5	3.40
		61.0	0.36
	sprinkler	30.5	11.10
		61.0	5.36
22.4	furrow	30.5	3.64
		61.0	1.37
	sprinkler	30.5	19.70
		61.0	5.90

Aldicarb residues leached deeper than one meter 6-9 months after application to a sandy Long Island soil (39). Based on a comparison of computer simulation modeling and ground water data, at least several percent of the aldicarb applied to certain Long Island fields leached to ground water (37).

Atrazine. Atrazine is a chlorinated, s-triazine herbicide. Its water solubility is 33 ppm (18), its vapor pressure is 3 X 10^{-7} torr (19), and H $<10^{-7}$ atm-m^3/mole. The pK_a of atrazine is 1.68 (40). K_{oc} values of 51 and 214 have been reported (19,20) and the mean K_{oc} of 56 soils was reported to be 163+80 (41). The K_d of atrazine varies between 1 and 8 in certain soils, and seems to be dependent on the cations present (42). K_d values of 0.40-0.46 were also recently reported for aquifer sands in Nebraska (43). Atrazine is in Helling's mobility class 3, which is the intermediate mobility class (10). (Class 5 is the most mobile, and class 1 is immobile.)

In the field, atrazine has been found to have a half-life of <1 month, but the half-life is affected by the tillage system (44-46), the agricultural soil ammendments and soil pH (45-48), and soil organic matter (49). In another study, atrazine and hydroxyatrazine have been found to persist into the following growing season (50).

Atrazine degraded in estuarine water with a half-life of 3-12 days, in two estuarine sediments with half-lives of 15 and 20 days, and in two agricultural soils with half-lives of 330 and 385 days; it was concluded that degradation is slower in agricultural soils than in estuarine systems (51). In two different soils under lab conditions, where the pHs were 4.8 and 6.5, respective half-lives of 53 days and 113 days were found (52).

In solution buffered at pH 5, the hydrolytic half-life decreased from 12 weeks to 6 weeks as the temperature increased from 20° to 30°C. Hydrolysis at pH 7 and greater proceeded with a half-life >200 days (52). In buffered solutions and in the presence of fulvic acid at a concentration of 0.5 mg/ml and at 25°C, the hydrolytic half-lives of atrazine were found to increase from 35 days to 742 days as the pH was increased from 2.9 to 7.0 (53). However, the half-lives decreased to about 10% of those values when the fulvic acid concentration was increased to 5 mg/ml.

Atrazine has been found in ground water in Nebraska (24,43, 56,57), Wisconsin (54), and Iowa (55). The concentrations in several dozen wells were typically 0.8 ppb. The four positives out of nine wells sampled in Iowa were associated with an adjacent river which was contaminated with atrazine by runoff. In 1979 and 1980, positive correlations between atrazine and nitrate in ground water were reported (56,24). It has been estimated that 0.07% of the atrazine applied to the surface leached to a 1.5 m depth in Nebraska (57).

Bromacil. This herbicide is a brominated uracil with an 815 ppm water solubility (18). Its vapor pressure is 8×10^{-4} mm Hg at 100°C (58). Bromacil is in mobility class 4 according to Helling's classification scheme (10). It has K_{oc} values ranging from 69-77 and K_d values of 0.2-1.8 in sandy loam soil and subsoil (20, 59).

Its half-life under field conditions is 5-6 months (60).

Bromacil leached to a shallow water table aquifer in a Lakeland sand soil in a Florida test plot (61). The concentrations seemed to be related to rainfall, and at one point exceeded 1 ppm, though values of several hundred parts per billion were more typical.

Carbofuran. Carbofuran is a benzofuranyl carbamate nematicide/ insecticide with reported water solubilities of 320 ppm (62), 415 ppm (20) and 700 ppm (19,58,63). Its vapor pressure is 8.3×10^{-6} torr, and its H value of 3.45×10^{-9} atm-m^3/mol is extremely low. Its K_d values are generally less than 2 and the mean K_{oc} for five soils was reported to be 29+9 (41). Another reference (64) showed the following K_d values for 7 soils; the % organic carbon is provided in the parentheses: 0.25(0.4); 0.74(1.2); 1.40(2.7); 1.13(3.1); 1.39(3.5); 2.22(7.5) and 8.74(16.8). Corresponding K_{oc} values range from 29.6-62.5.

Carbofuran has hydrolytic half-lives of about one year at pH 6, of about 4 weeks at pH 7, and less than one day at pH 9. Carbofuran phenol is the principal product formed at pH 5-9 (19).

The lab half-life of carbofuran in six soils (each under 4 different levels of moisture and 2-3 different temperatures) was in the range of 5-261 days. After 28 days of aerobic incubation in two of the six soils, parent carbofuran was the major extractable compound; degradation products comprised <5% of the extractable material (65). However, under anaerobic (flooded soil) conditions, the degradation product carbofuran phenol was found to be the principal product and to persist (66).

In the field, half-lives of 1-5 weeks were found in non-sandy soils containing 2.5-82% organic matter (67,68); however, as mentioned above (65), longer half-lives were found in lab studies where the soils contained 0.8-3.9% organic matter. There is evidence that fields previously treated with carbofuran degrade carbofuran more rapidly than fields not previously treated with carbofuran (67). However, a field study conducted in Minnesota (69) and a lab study using soil from a rice field (66) showed no difference in carbofuran degradation between previously treated and untreated plots. There is also evidence that carbofuran degrades more rapidly in soils treated with captafol (70).

Carbofuran has been found in ground water in Wisconsin and New York, in areas with sandy soils and water table aquifers, at levels typically between 1 and 50 ppb (19).

DBCP. The nematicide 1,2-dibromo-3-chloropropane has reported water solubilities of 700 ppm (72) and 1000 ppm (25). With a vapor pressure of 0.8 torr, H = 4 X 10^{-4} atm-m^3/mole.

K_d values can be derived from adsorption data in silt loam and in fine sand soils as 2.1 (73) and 0.13 (74), respectively. The corresponding K_{oc} values are 129 and 149, respectively. One can also estimate K_d and K_{oc} values from a field study in a clay loam soil of 0.28 and 55, respectively (75).

DBCP hydrolyzes via an E_2 mechanism to unsaturated products, with an estimated half-life of 141 years at pH 7 and 15°C, and 38 years at pH 7 and 25°C (72). In an amended, highly bioactive agricultural soil in the lab, DBCP was biodegraded over a period of several weeks (76).

DBCP has been found in ground water in Hawaii, California, Arizona, South Carolina, and Maryland (5-7,19,77-79). Typical positives are 0.02-20 ppb. Areas with the highest frequency of positives and the highest well concentrations are the San Joaquin Valley in California and the region southwest of Phoenix, Arizona. The Hawaii contamination has occurred despite several hundred feet of overburden between the basal aquifer and the surface. One set of California soil core results show that ppb amounts of DBCP has leached about 15 m through the unsaturated zone (6), whereas DBCP was not detected in another set of California soil cores sampled as deep as 10 m and five years after the last application (80). The latter results can possibly be explained by rapid movement of DBCP down the soil profile to depths greater than 10 m.

DCPA + Products. The parent compound is a chlorinated phthalate ester herbicide. Its vapor pressure is <1 X 10^{-2} torr at 40°C (18) and the vapor pressure of its acid degradation products are unknown but most likely lower. DCPA water solubility is only 0.5 ppm (18), but its acid products likely have much higher water solubilities. The average half-life of the parent is 100 days in most general soil types (18) but half-lives of 18-37 days with mono and diacid products have also been reported (19). With regard to contamination of ground water, the DCPA residues of concern are the acid degradation products, rather than the parent compound.

In a 1979-1980 Long Island, New York monitoring study, Diamond Shamrock Corp. found DCPA residues (parent + monoacid + diacid) in 6 wells out of about 10 sampled in an agricultural area in Suffolk County, New York. (71). Typical positives were 40-600 ppb and sampling was conducted monthly over a one-year period.

Dichloropropane. 1,2-Dichloropropane (DCP) is in the same class of soil fumigant nematicides as DBCP and EDB (1,2-dibromoethane). Its water solubility is 2,600-2,700 ppm (81), its vapor pressure is 42 torr (82), and the calculated value of H is

2.1 X 10^{-3} atm-m^3/mol. Therefore, it is more volatile from water than any other pesticide discussed in this paper.

DCP has a K_d of 0.43 and a K_{oc} of 46 in a silt loam soil (73). It will sorb to clay minerals in dry soils but will desorb as the soil absorbs moisture (83). It also sorbs to organic matter (84). However, areas where nematicidal fumigants are principally used, such as California, and the coastal plain areas of Georgia, South Carolina, North Carolina and Virginia (85) generally have sandy soils and low organic carbon content and would probably have little impact on reducing mobility of DCP by soil adsorption.

Based on the half-lives of one and two carbon chloroaliphatics, the hydrolytic half-life of DCP is expected to be 6 months to several years (86). Vapor phase photolysis under simulated sunlight did not occur after prolonged exposure (87).

It readily volatilizes from soil, but is stable in sandy loam soil when incubated in a closed container for 12 weeks (88). In a medium loam soil, some (5%) radioactivity was unextractable after 20 weeks of incubation of [2-^{14}C]1,2-dichloropropane (88).

DCP has been found in ground water in about 3 Maryland wells (7), about 30 Long Island, New York wells (37), and over 60 California wells (89) at levels typically ranging between 1 ppb and 50 ppb. Soil core data from California showed that DCP was detected as far as 7 m down at 0.2-2.2 ppb (89).

Dinoseb. Dinoseb is a dinitrophenol herbicide with a pK_a of 4.35 (90) and a water solubility of 52 ppm (18). Its vapor pressure is 5 X 10^{-5} torr (91) and H is 3.0 X 10^{-7} atm-m^3/mol.

Dinoseb has a reported K_{oc} of 124 (20) but no K_d was given. The adsorption of dinoseb on montmorilonite and illite clays with increasing pH resembles its pH-titration curve; adsorption decreases with increasing pH until there is little, if any, adsorption when the pesticide is in its anion form (90).

Dinoseb can leach in certain soils (18) and microbial breakdown has been demonstrated in soil (18) and in pure culture (92).

Dinoseb has been found in Long Island, New York, in 6 wells out of 66 sampled, at levels of 4.5 ppb and less (93).

EDB. 1,2-Dibromoethane is structurally very similar to DBCP and has similar properties and use patterns. Its water solubility is 4,300 ppm (25) and its vapor pressure is 11 torr (94). H is 6.33 X 10^{-4} atm-m^3/mol.

The K_d value for EDB in silt loam soils has been reported as 0.25 (95) and 0.6 (73), with K_{oc} values ranging from 36-158. There is also a report of K_d/K_{oc} values in two unspecified soils of 0.408/39.7 and 0.803/40.0, respectively (96).

One study (76) screened about 100 soils for bioactivity and chose one of the more bioactive to study the soil metabolism of EDB. Approximate half-lives were calculated from the results of this study and, even in a single agricultural soil, they varied from 1.5-18 weeks in different soil aliquots of that soil.

A study is in progress which is investigating the kinetics and mechanism of hydrolysis of EDB at pH 5, 7, and 9, and three elevated temperatures (97). Preliminary indications are that EDB's half-life at ambient ground water temperatures and pH is >6 years, and the most probable hydrolysis product is vinyl bromide.

EDB has been found in ground water in Florida, Georgia, California, South Carolina, and Hawaii, although it is not clear at this time whether the Hawaii sites are due to normal agricultural use (98-100). Typical positives are 0.05-20 ppb. Information from South Carolina indicates that EDB has been found at levels of 0.036-0.24 ppb in 3 out of 19 wells sampled in the southcentral part of the state (100). The sampling programs in progress in these states are likely to be expanded to other states. The Georgia findings in Seminole County are particularly disturbing because the EDB apparently leached through a confining layer to an artesian aquifer (98,101). California soil core data show that EDB leached about 10 m at the ppb level (80). The soil core result was one of the driving factors behind EPA's recent emergency suspension of soil fumigation uses of EDB (98).

Oxamyl. Oxamyl is an oxime carbamate with a 280,000 ppm water solubility (63) and a 2.3×10^{-4} torr vapor pressure (19). It is used as an insecticide/nematicide.

Oxamyl is entirely stable to hydrolysis for 11 days at pH 4.7 and room temperature, has a half-life of about 6 days at pH 6.9, and an estimated half-life of less than 1 day at pH 9.1 (102). Degradation is more rapid in river water (102).

Lab half-lives in four non-sandy soils (15°C, pH 7.18-7.95 and moisture content of 18-25%) were 6-18 days but in a sandy loam soil, the following results were found (27).

Table II.   Oxamyl Degradation in a Sandy Loam Soil Under Lab Conditions (27)

Temperature (°C)	Moisture (% by mass)	Half-life (days)
5	10	33-58
10	10	18-31
15	5	25-28
15	10	16-21
15	15	16-22

Based on adsorption data from a sandy loam soil with organic matter content varying from 0.81% to 5.92% organic matter K_d ranges from 0.02 to 0.3, and K_{oc} values ranging from less than 0.2 to 8.6 can be calculated (103). This range in K_{oc} is wide; refer to the Discussion section for further elaboration on this.

Another lab study (102) using a loamy sand (pH 6.8) and a fine sand (pH 6.4) showed oxamyl half-lives of 11 and 15 days under aerobic conditions, respectively. There were no extractable degradation products. However, under anaerobic conditions, a half-life of 6 days was found and extractable degradation products persisted (102).

Under field conditions, 10% of the applied oxamyl remained after 2 months (103) which is generally consistent with the data above.

Oxamyl has been found in Long Island, New York ground water at concentrations typically between 5 and 65 ppb (19).

Simazine. Simazine is a triazine herbicide which contains one less methyl group than atrazine, yet its water solubility of 3.5 ppm (18) is one-tenth of atrazine's. Its vapor pressure is 6.1×10^{-9} torr (104). The pK_a of simazine is 1.65 (40). Its average K_{oc} for 147 soils was reported as 138 ± 17 and its K_d is typically less than 5 (41). Using Helling's data for 14 soils (105), K_ds from 0.26 to 5 were calculated. Simazine is in mobility class 3 (intermediate) according to Helling's scheme (44).

In solution buffered at pH 5, the hydrolysis half-life decreased from 12 weeks to 8 weeks as the temperature increased from 20° to 30°C. Hydrolysis at pH 7 and greater proceeded with a half-life >200 days (52).

In two different soils under lab conditions, where the pHs were 4.8 and 6.5, respective halflives of 45 days and 113 days were found (52). The half-life under field conditions is <1 month but hydroxysimazine, the major degradation product, may persist (106). The more acidic the soil, the more rapid is the hydrolytic reaction forming hydroxysimazine (107). Persistence of simazine increases as soil pH increases, with maximum persistence at a pH of 6.6, and it is affected by the method of tillage (108).

In a recent California study, designed for even spatial distribution, simazine was found in 6 wells out of 166 sampled at levels between 0.5 ppb and 3.5 ppb (99). In a different part of the same study, simazine was found throughout the soil profile down to ground water which was encountered at about 9 m (80). The estimate made was that 8.2% of the simazine applied over an 11 year period remained in the soil column above the aquifer.

The ground water monitoring results for all pesticides are summarized in Table III.

Table III.   TYPICAL POSITIVE RESULTS OF PESTICIDE GROUND WATER
             MONITORING IN THE U.S.

Pesticide	Chemical Class	State(s)	Typical Positive, ppb
Alachlor	acetanilide	NB	0.04
Aldicarb (sulfoxide & sulfone)	carbamate	AZ, CA, FL, ME, MO, NC, NJ, NY, OR, TX, VA, WA, WI	1-50
Atrazine	triazine	IA, NB, WI	0.3-3
Bromacil	uracil	FL	300
Carbofuran	carbamate	NY, WI	1-5
DBCP	low M.W.[a] halogenated hydrocarbon	AZ, CA, HI, MD, SC	0.02-20
DCPA (and acid products)	phthalate	NY	50-700
1,2-Dichloro-propane	low M.W. halogenated hydrocarbon	CA, MD, NY	1-50
Dinoseb	dinitrophenol	NY	1-5
EDB	low M.W. halogenated hydrocarbon	CA, FL, GA, SC	0.05-20
Oxamyl	carbamate	NY	5-65
Simazine	triazine	CA	1-2

[a] M.W. = molecular weight

There may be other pesticide contaminants and ground water contamination sites yet to be discovered. Therefore, the 12 different pesticides found in a total of 18 different states, as listed in Table III, represents the minimum number of pesticides and their distribution. There have been several reports of picloram in well water, most of them anecdotal. However, it cannot be substantiated that the contamination has been due to normal pesticide use and leaching through the soil or due to spillage.

The key results of the environmental transport and persistence analyses for these 12 pesticides are summarized in Table IV. The processes included in soil dissipation half-lives could be biodegradation, hydrolysis, oxidation, reduction, soil photolysis, and volatilization. Note that biodegradation may be other than first-order, whereas the other processes are usually first order or pseudo-first-order.

The data in Table IV can be used to draw the general conclusions given at the end of this report.

Mathematical Modeling

Programs that model the transport and persistence of pesticides in the subsurface soil environment have been developed (39,109, 110) and are continually being improved. An interactive, one-dimensional model being used is PESTANS, (109). It can be readily used in the early stages of the hazard assessment process. One of the principal drawbacks of this analytic solution is that throughout the entire zone of leaching there is only provision for one rate constant and one K_d value. In the field, these parameters vary with depth as the soil profile changes from surface to subsurface. Also, the rate may change due to differences in dissolved oxygen content and microbial populations. This limitation could be overcome with a slight modification in the computer program. Another drawback is that an average recharge rate is used which the user usually calculates from annual data. This can result in an underestimate of true short term leaching rates especially at times of high water flux.

A successor to PESTANS has recently been developed which allows the user to vary transformation rate and K_d with depth; i.e., it can describe nonhomogeneous (layered) systems (39,111). This successor actually consists of two models - one for transient water flow and one for solute transport. Consequently, much more input data and CPU time are required to run this two-dimensional (vertical section), numerical solution. The model assumes Langmuir or Freundlich sorption and first-order kinetics referenced to liquid and/or solid phases, and has been evaluated with data from an aldicarb-contaminated site in Long Island. Additional verification is in progress. Because of its complexity, it would be more appropriate to use this model in a higher level, rather than a screening level, of hazard assessment.

TABLE IV. KEY ENVIRONMENTAL FATE PARAMETERS FOR PESTICIDES FOUND IN GROUND WATER[a]

Pesticide	K_{oc}[b]	K_d	Solubility (ppm)	Hydrolysis half-life (weeks)	Soil/field dissipation[c] half-life (wks)	pK_a
Alachlor	213[d]	0.6-8.1	240	NA	1-10	
Aldicarb (sulfoxide & sulfone)	parent-36[d] sulfox.- 42; 405	parent<4 sulfox.-3.3, 0.34	6000 - 43000	10-650[e]	about 10	
Atrazine	51; 163+80 638[d]	0.4-8	33	10-106	<4-57	1.68
Bromacil	109[d]; 69-77	0.2-1.8	815	NA	20-26	
Carbofuran	30-62; 29+9 120[d]; 160[d]	0.25-8.7	700; 415	2-50	1-37	
DBCP	55; 149; 119[d]; 129	0.13; 2.1; 0.28	700; 1000	2000-7300	about 10[g]	
DCPA (and acid products)	NA	NA	NA	NA	NA	<3.5[f]

1,2-Dichloro-propane	46; 58d	0.43	2,600	25-200 (estimate)	NA	4.35
Dinoseb	124; 497d	NA	52	NA	NA	NA
EDB	36-158; 44d	0.25-0.80	4300	>300	2-18	
Oxamyl	0.2-8.6; 4.4d	0.02-0.3	280,000	<1-few wks.	1-8	
Simazine	138; 2700d	0.26-5	3.5	8-30	<4-16	1.65

a/ These data were measured at or extrapolated to ambient temperature and pH values. The data are discussed in the text. NA = not available. b/ K_{oc} = soil water distribution coefficient (K_d) divided by the organic carbon content of the soil. c/ Whenever possible, half-life for soil dissipation is derived from the field data half-lives described in the text rather than lab data. As such, it may not represent a true first-order process. d/ Value has been estimated from the equation in ref. 20. e/ Hydrolysis of total residues (aldicarb + sulfoxide + sulfone). f/ pKa for p-phthalic acid is 3.5. The chlorine atoms of DCPA should lower the pKa to about 2. g/ Conditions optimized for soil metabolism.

A continuous, dynamic, one-dimensional model called the Pesticide Root Zone Model or PRZM, has been developed recently by EPA/ORD in Athens, Georgia (110). PRZM allows for varying hydrologic and chemical properties by soil horizon. Weather data for water flow modeling is obtained from daily precipitation records of the National Weather Service. It has been successfully validated with atrazine field data from Watkinsville, Georgia and aldicarb data from Long Island, New York for depths less than 3 meters.

PRZM was applied to a hypothetical situation of a pesticide in a Georgia agricultural environment. An overall, pseudo-first-order degradation rate coefficient of 0.001 day^{-1} was used, along with a series of K_d values. A cover crop of peanuts was assumed. The simulation was done for a 900 g/ha application to a class A soil (well drained) and a class D soil (poorly drained). Movement through the root zone was simulated using rainfall records. In the hypothetical 1-ha plot, 800 g and 550 g of the pesticide leached past 60 cm in the class A and D soils, respectively, when a K_d value of 0.06 was used; 40 g and 5 g leached past 60 cm in the class A and D soils, respectively, when a K_d value of 1.5 was used. These computational results support the conclusion on K_d values stated at the end of this paper.

Mathematical modeling of subsurface pesticide fate is gaining increasing acceptance among EPA and industry scientists. These models can be used in many ways. The pesticide industry can use models in the early stages of product development for potential hazard screening. EPA and pesticide scientists can use models to design monitoring studies, as well as to amend product labeling if confirmatory field data exist. Finally, at some point in the future, modeling results, coupled with toxicity concerns, may form the principal bases for major regulatory actions.

Discussion

In this section are more detailed discussions about soil sorption, volatility, soil core data and photolysis data and how they bear on leaching potential.

Sorption. K_{oc} values have been looked upon as constant for neutral organic chemicals, independent of soil type (73). The parameter is valid for comparison of leaching potential of pesticides with widely varying water solubility; however, it is difficult to make quantitative comparisons among polar organics. For example, note the wide range in K_{oc} for oxamyl described in Table IV. This may be due in part to interactions with soil mineral fractions which become important when the soil organic matter content is low (56,112). Also, use of water solubility as a predictor of K_{oc} (20,73) can be misleading. For example, simazine's water solubility is only 3.5 ppm and one would predict a K_{oc} of 2,700 on the basis of one of the several correlation equations in the literature (20), yet its mean K_{oc} is actually 138+17. Following is a new interpretation of pesticide sorption which may explain this phenomenon. Its basis is an interpretation of ring electrical effects.

It is generally recognized that heteroaromatic compounds, such as atrazine and simazine, contain nitrogen atoms bearing partially localized electron density (i.e., partial negative charges) (113). A diagram can be constructed which displays this concept for the triazine system. Hydrogen atoms have been eliminated from the diagram below, but all would possess a partial positive charge. The symbol delta (δ) indicates a partial charge.

(The exocyclic ring nitrogens probably conjugate with the triazine ring system, affecting the charge picture to a small extent.) If this charge picture is correct, then the solvent water molecules do not "see" a very polar molecule, i.e., a molecule with significant positive and negative charges at opposite ends. Instead, the solvent water molecules "see" a rather balanced charge system. This would explain the low water solubilities of atrazine and simazine (<<100 ppm). So why are these molecules not tightly bound to soils? The answer, again, may lie in the charge picture. Most clays are layer silicates, and the montmorillonite type undergoes isomorphous substitution of magnesium for some of the aluminum in the clay lattice, resulting in a net negative charge. Cations, such as Ca++, H+, Na+, and K+, can be adsorbed to the surface of the clay particles (114). When cations with low positive charge densities such as Na+ and K+ are present, the negative charges of the micellar surfaces can be more available for interaction with solutes such as the triazines. However, referring back to the triazine figure above, we see that the outer part of the molecule, as represented by the plane of the paper, could have most of the partial negative charges, whereas the partial positive charges are somewhat shielded by the chlorine atom and the neutral organic groups. Also, note that aromatic rings, such as triazines, contain significant pi electron density above and below the plane of the ring which may interact with the clay surface. Thus, negative charges on the clay and partial negative charges on the pesticide would repel one another resulting in no net attraction; hence, minimizing adsorption. The basic concept behind this theory is not new (90).

This theory of charge repulsion could explain why K_{oc}s for simazine and atrazine, and likely other triazines, are so much lower than that predicted by their water solubilities. The same theory may apply to other classes of compounds, such as carbamates and aromatic acids; further investigation is warranted.

Volatility. When EPA began emphasizing work in pesticide contamination of ground water in 1979, it seemed reasonable to assume that volatility would be an important factor for determining loading to the subsurface environment. It is clear that the use of vapor pressure alone would do little to focus a list of potential ground water contaminants. For example, 1,2-dichloropropane (DCP) has a vapor pressure of about 40 torr at 20°C, over twice that of water, and higher than most pesticides, yet it is an agricultural pollutant in dozens of wells across the U.S. Regarding nematicides, the volatility must be low enough in order for the nematicide to penetrate well into the root zone and persist long enough to effectively decrease the nematode population. (Note that use practices, e.g., soil incorporation of DCP, possibly followed by irrigation water, tend to decrease vapor losses). A more appropriate volatility criterion is probably the Henry's law constant (H), as defined earlier. The data presented here for these ground water contaminants show that H does not exceed 2.1×10^{-3} atm-m^3/mol at ambient temperatures, which makes these compounds less volatile in aqueous systems than benzene and carbon tetrachloride (11). One may ask, what is the significance of aqueous volatility when the concern is soil losses? The answer is that (1) these chemicals are often applied in irrigated fields or shortly preceding rainfall, and (2) work by Dow scientists has shown that vapor pressure, water solubility, and K_{oc} are the determining factors in volatilization from the soil (90).

There is also another pertinent question to be asked – what is the relevance of these estimated H values, with their awkward units, to the natural environment? One answer is that comparisons of H values between different neutral organics can be helpful in a relative sense, particularly when one or more of the chemicals in the comparison have already been well characterized in the field. A second answer is that H values can be used in calculations of volatilization from rivers (11) and soil surfaces (91). Finally, to put H values in further perspective, the following soil volatilization data have been obtained from the literature (Table V). Clearly, the data are inadequate to allow direct correlation between H and volatilization losses from soil. The data do show, however, that (1) even compounds with low H values can volatilize when applied to the surface, and (2) differences in application methods may significantly affect volatilization losses.

Soil Cores. Monitoring of ground water for contamination by pesticides must be accompanied by soil core sampling and analyses. The absence of pesticide residues in ground water for several years after pesticide application does not necessarily mean the pesticide has degraded in the soil. It could mean that the pesticide residues are stable in soil and are slowly leaching, but have not yet reached the water table. Therefore, a determination as to whether the potential for ground water contamination due to a pesticide is real cannot be made without the results of soil core analyses. It is important to note that soil core analyses can be meaningless unless adequate precautions are taken to ensure that deeper cores are not contaminated by higher concentrations of pesticides from the surface during the sampling process and that other sources of contamination are minimized or eliminated.

Photolysis. Photolysis is another mode of pesticide degradation having impact on persistence of the pesticide and on which degradation products will be form. Although many pesticides are soil incorporated or watered–in, thereby diminishing potential for degradation by photolysis, some of the pesticide will remain on the surface and be subjected to photolysis. In addition, runoff water from pesticide treated fields is often collected and impounded in tailwater pits for reuse as irrigation water. Such tailwater pits in Kansas have been found to contain alachlor, atrazine and carbofuran, among other pesticides (115). Atrazine in tailwater pits from Nebraska has been found to leach through the bottom of the pit to ground water (116). The pesticides in tailwater pits would be subject to photolysis before being reapplied to the fields during irrigation. Photooxidation, being a major mode of photodecomposition of pesticides (117,118), could result in the formation of transformation products more toxic and more leachable than the parent compound.

Table V. Comparing Volatilization Results and H Values

Chemical	Volatilization Results	H (atm-m³/mol)	Conditions	Ref.
carbofuran	24 day half-life	3.4×10^{-9}	lab – surface applied airspeed 1 km/hr, moist loam, 25°C.	91
dinoseb	26 day half-life	3.0×10^{-7}	"	"
atrazine	45 day half-life	2.0×10^{-6}	"	"
EDB	less than 5% lost by diffusion to the surface	6.3×10^{-4}	lab – injected @ 30.5 cm into soil and covered, 0.8 km/hr airspeed, sandy loam, 15°C & 25°C.	84
DCP	less than 5-10% lost by diffusion to the surface	2.1×10^{-3}	"	"

Unfortunately, photolysis data for pesticides submitted to the Agency before 1980 are often inadequate. For ground water assessments, photolysis on soil surfaces is a concern and, to some extent, in natural waters such as tailwater pits as noted above. However, much of the photolysis data on file with the Agency involve studies conducted in methanol or hexane, on glass plates, at an unrealistically high solute concentration, or under wavelengths unrepresentative of natural sunlight. Current EPA pesticide guidelines (119) specify the testing to be done under conditions more representative of those found in the environment. However, extrapolating laboratory results to a field situation is complicated by the fact that the environment contains naturally occurring photosensitizers such as humic acids and iron (III) (120,121) but the Guidelines (119) allows the Agency to address the problem on a case-by-case basis. These triplet sensitizers can induce pesticides to react by significantly increasing the population of low lying triplet states; pesticides ordinarily do not react from the higher energy, shorter lived singlet states. Photolysis may be a parameter of concern, but due to the uncertainties in the data, photochemical assessments have not been included in the capsule chemical summaries.

Conclusions

It is often difficult to evaluate dissipation and leaching data in that varying field and laboratory test conditions can have significant impact on the test results. In addition, the use of different test protocols by investigators is not always amenable to comparison of test results. Also, there are limitations in doing computational/modeling evaluations because of lack of information on environmental parameters at the sites of concern or because of necessary oversimplifications by the models. (However, meaningful environmental parameters representative of reasonable worst case conditions can usually be selected). Also, the authors have been able to identify only a few references in the published literature which include the results of actual field monitoring. In light of the potential problem posed by pesticide contamination of ground water, it is clear that much more ground water monitoring must be done. Nonetheless, the information currently available suggests certain generalizations which may provide useful guidance when making hazard assessments. Specifically, based upon the limited data available, it appears that when all the following chemical characteristics and field conditions appear in combination, the potential for ground water contamination is high. Therefore, if all or most of these characteristics are met for a new pesticide/use pattern being researched by a pesticide registrant, computer modeling and field testing should be considered to ascertain actual potential for ground water contamination.

Pesticide Characteristics

- o Water solubility – greater than about 30 ppm. See the discussion above for the potential importance of electronic interactions to this criterion.

- o K_d – less than 5, and usually less than 1 or 2.

- o K_{oc} – less than 300–500.

- o Henry's law constant – less than 10^{-2} atm-m^3/mol

- o Speciation – negatively charged (either fully or partially) at ambient pH.

- o Hydrolysis half-life – greater than about 25 weeks.

- o Photolysis half-life – greater than about 1 week (but this criterion may not be as influential as the others. See photolysis discussion above).

- o Soil half-life – greater than about 2–3 weeks.

Field Conditions

- o Recharge – total precipitation and irrigation recharge greater than about 25 cm/yr. An important factor in this criterion is the soil's drainage ability; i.e. soils with low moisture holding capacity are conducive to high recharge.

- o Nitrates – high levels in the ground water are indicative of pesticide ground water contamination potential.

- o Aquifer – unconfined; porous soil above unconfined aquifer.

- o Soil with a pH providing high stability to the pesticide residues.

Although these generalizations were not intended to be applied to waste disposal sites, the generalizations likely have some relevance to uncontrolled hazardous waste sites, as long as the reader remembers the differences described in the beginning of the Chemical Characteristics section.

As seen from Table III, two-thirds of the pesticides found in ground water from agricultural uses fall in one of the following three categories: carbamates, triazines, and low molecular weight halogenated hydrocarbons. It can be expected that another class which will be found in this evolving field is aromatic acids, as represented by the pyridine acids and the chlorinated benzoic acids. This is because, based on a cursory review of the literature, the properties of this class of compounds seem to meet the criteria for pesticide ground water contaminants described.

Finally, an equally important component of ground water risk assessment is toxicity. Only rarely have levels of pesticides in well water been detected which would cause acute toxicity, unless improper disposal caused the contamination. Rather, as can be seen in Table III, the pesticide levels are usually in the low ppb range. Therefore, our current toxicity concerns are usually for chronic human toxicity or, occasionally, aquatic toxicity. There is also the possibility of organisms receiving toxic amounts of pesticide residues over time via biomagnification.

Legend of chemical names and abbreviations

Alachlor 2-Chloro-2',6'-diethyl-N-(methoxymethyl)acetanilide
Aldicarb 2-Methyl-2-(methylthio)propionaldehyde O-(methyl-
          carbamoyl)oxime
Aldicarb sulfoxide 2-Methyl-2-(methylsulfinyl)propionaldehyde
          O-(methylcarbamoyl)oxime
Aldicarb sulfone 2-Methyl-2-(methylsulfonyl)propionaldehyde
          O-(methylcarbamoyl)oxime
Atrazine 2-Chloro-4-(ethylamino)-6-(isopropylamino)-s-triazine
Bromacil 5-Bromo-3-sec-butyl-6-methyluracil
Captafol cis-N-[(1,1,2,2-Tetrachloroethyl)thio]-4-cyclohexene-
          1,2-dicarboximide
Carbofuran 2,3-Dihydro-2,2-dimethyl-7-benzofuranyl methylcarb-
          amate
Carbofuran phenol 2,3-Dihydro-2,2-dimethyl-7-hydroxybenzofuran
DBCP 1,2-Dibromo-3-chloropropane
DCPA Dimethyl tetrachloroterephthalate
Dichloropropane 1,2-Dichloropropane
Dinoseb 2-sec-Butyl-4,6-dinitrophenol
EDB 1,2-Dibromoethane
Oxamyl Methyl N',N'-dimethyl-N-[(methylcarbamoyl)oxy]-1-thio-
          oxamimidate
Simazine 2-Chloro-4,6-bis(ethylamino)-s-triazine

AZ	Arizona	NC	North Carolina
CA	California	NJ	New Jersey
FL	Florida	NY	New York
GA	Georgia	OR	Oregon
HI	Hawaii	SC	South Carolina
IA	Iowa	TX	Texas
MD	Maryland	VA	Virginia
ME	Maine	WA	Washington
MO	Missouri	WI	Wisconsin
NB	Nebraska		

Literature Cited

1. "Pesticide Industry Sales and Usage - 1982 Market Estimates," U.S. Environmental Protection Agency, 1982.
2. Knusli, E., in "Advances in Pesticide Science, Part I";
   Pergamon Press: Oxford, England, 1979.
3. McEwen, F.L. and Stephenson, G.R. "The Use and Significance of Pesticides in the Environment"; John Wiley and
   Sons: New York, 1979; p.471.
4. "The Report to Congress: Waste Disposal Practices and
   Their Effects on Ground Water," U.S. Environmental Protection Agency, 1977.
5. Peoples, S. A.; K.T. Maddy, W. Cusick, T. Jackson, C. Cooper
   and A.S. Frederickson; Bull. Environ. Contam. Toxicol.
   1980, 24, 611-618.
6. Nelson, S.; Iskander, M.; Volz, M.; Khalifa, S.; Haberman,
   R. Sci. Total Environ. 1981, 21, 35-40.
7. Pinto, E. "Report of Groundwater Contamination Study in
   Wicomico, Maryland"; Wicomico County Health Dept.: State of
   Maryland, 1980 (plus 1981 - 1982 addenda).
8. Zaki, M.H., Moran, D.; Harris D. Am. J. Public Health 1982,
   72, 1391-1395.
9. Rothschild, E. R.; Manser, R. J.; Anderson, M. P. Ground
   Water 1982, 20, 437-445.
10. Helling, C.S. Soil Sci. Soc. Am. Proc. 1971, 35, 737-743.
11. Mackay, D.; Yuen, T.K. Water Pollu. Res. J. Canada 1980,
    15, 83-98.
12. Rao, P.S.C.; Davidson, J.M. Water Research 1979, 13,
    375-380.
13. Davidson, J.M.; Rao, P.S.C.; Ou, L.T.; Wheeler, W.B.;
    Rothwell, D.F. " Adsorption, Movement, and Biological
    Degradation of Large Concentrations of Selected Pesticides
    in Soils," 1980, EPA-600/2-80-124.
14. Ou, L.-T.; Rothwell, D.F.; Wheeler, W.B.; Davidson, J.M.
    J. Environ. Qual. 1978, 7, 241-246.
15. Ou, L.-T.; Davidson, J.M.; Rothwell, D.F. Soil Biol. Bio-
    chem. 1978, 10, 443-445.
16. Rao, P.S.C.; Hornsby, A.G.; Nkedi-Kizza, P. Am. Chem. Soc.
    186th Nat. Mtg., Environ. Chem. Div. Abstracts, 1983.
17. Ghassemi, M.; Quinlivan, S.; Haro, M.; Metzger, J.;
    Scinto, L.; White, H. "Compilation of Hazardous Waste
    Leachate Data", 1983, Office of Solid Waste, U.S. Environ-
    mental Protection Agency, EPA Contract No. 68-02-3174.
18. "Herbicide Handbook," 5th edition, Weed Sci. Soc. Amer.,
    Champaign, IL, 1983; 515 pages.
19. Data summarized from EPA pesticide registration files,
    unpublished.

20. Kenaga, E.E.; Goring, C.A.I. in "Relationship Between Water Solubility, Soil Sorption, Octanol-Water Partitioning, and Concentration of Chemicals in Biota," Aquatic Toxicology, ASTM STP 707, Eaton, J.G.; Parrish, P.R.; Henricks, A.C., Eds.; American Society for Testing and Materials, 1980; pp. 78–115. (The K_{oc} data and the K_{oc} regression equation are also given in Kenaga, E.E. Ecotoxicol. Environ. Saf. 1980, 4, 26–38).

21. Stewart B.A.; Wollhiser, D.A.; Wischmeier, W.H.; Caro, J.H.; Frere, M.H. "Control of Water Pollution from Cropland: Volume I. A Manual for Guideline Development", 1975, Agricultural Research Service, U.S. Dept. of Agriculture, EPA-600/2-75-026a; USDA ARS-H-5-1.

22. "CREAMS: A Field Scale Model for Chemicals, Runoff, and Erosion from Agricultural Management Systems,"; Knisel, W.G.Jr., Conservation Research Report #26, USDA, 1980, 643 pages.

23. Yu, C.; Booth, G.M.; Hansen, D.J.; Larsen, J.R. J. Agric. Food Chem. 1975, 23, 877–879.

24. Spalding, R.F.; Junk, G.A.; Richard, J.J.; Pestic. Monit. J. 1980, 14 (2), 70–73.

25. Gunther, F.A.; Westlake, W.E., Jaglan, P.S. Residue Rev. 20, pp. 1–148.

26. Dr. Robert Bertwell, Union Carbide Corporation, personnal communication.

27. Bromilow, R.H.; Baker, R.J.; Freeman, M.A.H.; Gorog, K. Pestic. Sci. 1980, 11, 371–378.

28. Hough, A.; Thomason, I.J.; Farmer, W.J. J. Nematol. 1975, 7, 214–221.

29. Hansen, J.L.; Spiegel, M.H. Environ. Toxicol. Chem. 1983, 2, 147–153.

30. Bull, D.L. J. Econ. Entomol. 1968, 61, 1598–1602.

31. Bull, D.L.; Stokes, R.A.; Coppedge, J.R.; Ridgway, R.L. J. Econ. Entomol. 1970, 63, 1283–1289.

32. Andrawes, N.R.; Bagley, W.P.; Herrett, R.A. J. Agric. Food Chem. 1971, 19, 727–730.

33. Walter-Echols, G.; Lichtenstein, E.P. J. Econ. Entomol. 1977, 70, 505–509.

34. Timms, P.; MacRae, I.C. Bull. Environ. Contam. Toxicol. 1983, 31, 112–115.

35. Zinder, S.H.; Brock, T.D. J. Gen. Microbiol. 1978, 105, 335–342.

36. De Bont, J.A.M.; Van Dijken, J.P.; Harder, W. J. Gen. Microbiol. 1981, 127, 315–323.

37. Baier, J.H.; Robbins, S.F. "Report on the Occurrence and Movement of Agricultural Chemicals in Groundwater: South Fork of Suffolk County", Suffolk County Dept. Health Services, Hauppauge, New York 11788, Sept. 1982, pp. 50–51.

38. Elgind, G.M.; Van Gundy, S.D.; Small, R.H. Rev. Nematol.
     1978, 1, 207-215.
39. Enfield, C.G.; Carsel, R.F.; Phan, T. Quality of Ground-
     water, Proc. Int. Symp. 1981, p. 507.
40. Weber, J.B. Residue Rev. 1970, 32, 93-130.
41. Rao, P.S.C.; Davidson, J.M. in "Estimation of Pesticide
     Retention and Transformation Parameters Required Required
     in Non-Point Source Models"; Overcash, M.R.; Davidson,
     J.M., Eds.; Ann Arbor Science Publishers, Inc., 1980,
     pp. 23-67.
42. "A Literature Survey of Benchmark Pesticides," The George
     Washington University Medical Center, Washington D.C.,
     1976.
43. Wehtje, G.R.; Spaulding, R.F.; Burnside, O.C.; Lowry, S.R.;
     Leavitt, J.R.C. Weed Sci. 1983, 31, 610-618.
44. Bauman, T.T.; Ross, M.A. Weed Sci. 1983, 31, 423-426.
45. Lowder, S.W.; Weber, J.B. Weed Sci. 1982, 30, 273-280.
46. Kells, J.J.; Rieck, C.E.; Blevins, R.L.; Muir, W.M. Weed
     Sci. 1980, 28, 101-104.
47. Hiltbold, A.E.; Buchanan, G.A. Weed Sci. 1977, 25, 515-520.
48. Hance, R.J. Pestic. Sci. 1979, 10, 83-86.
49. Rahman, A.; Matthews, L.J. Weed Sci. 1979, 27, 158-161.
50. Khan, S.U.; Marriage, P.B.; Hamill, A.S. J. Agric. Food
     Chem. 1981, 29, 216-219.
51. Jones, T.W.; Kemp, W.M.; Stevenson, J.C.; Means, J.C.
     J. Environ. Qual. 1982, 11, 632-637.
52. Burkhard, N.; Guth, J.A. Pestic. Sci. 1981, 12, 45-52.
53. Khan, S.U. Pestic. Sci. 1978, 9, 39-43.
54. Collaborative project between the Wisconsin Department of
     Natural Resources, Madison, WI and EPA's Beltsville, MD
     pesticide laboratory, 1982-1983.
55. Richard, J.J.; Junk, G.A.; Avery, M.J.; Nehring, N.L.;
     Fritz, J.S.; Svec, H.J. Pestic. Monit. J. 1975, 9,
     117-123.
56. Spalding, R.F., Exner, M.E.; Sullivan, J.J.; Sullivan, P.A.
     J. Environ. Qual. 1979, 8, 374-383.
57. Wehtje, G.; Leavitt, J.R.C.; Spalding, R.F.; Mielke, L.N.;
     Schepers, J.S. Sci. Total Environ. 1981, 21, 47-51.
58. "Pesticide Manual," British Crop Protection Council, 1977,
     5th edition, Worcestershire WR9 OHX, England.
59. Leistra, M.; Frissel, M.J., in "Pesticides"; Coulston, F.;
     Korte, F., Eds.; Georg Thieme Publishers Stuttgart, 1975;
     pp. 817-828.
60. Gardiner, J.A.; Rhodes, R.C.; Adams, J.B.; Soboczenski,
     E.J. J. Agri. Food Chem. 1969, 17, 980-986.
61. Hebb, E.A.; Wheeler, W.B. J. Environ. Qual. 1978, 7,
     598-601.
62. Bowman, B.T.; Sans, W.W. J. Environ. Sci. Health 1979, 6,
     625.

63. "Farm Chemicals Handbook," Berg, G.L., Ed.; Meister Publishing Co., Willoughby, Ohio, 1981; p. C 248.
64. Felsot, A.; Wilson, J. Bull. Environ. Contam. Toxicol. 1980, 24, 778.
65. Ou, L.-T.; Gancarz, D.H.; Wheeler, W.B.; Rao, P.S.C.; Davidson, J.M. J. Environ. Qual. 1982, 11, 293-298.
66. Venkateswarlu, K.; Sethunathan, N. J. Agric. Food Chem. 1978, 26, 1148-1151.
67. Greenhalgh, R.; Belanger, A. J. Agric. Food Chem. 1981, 29, 231-235.
68. Gorder, G.W.; Dahm, P.A.; Tollefson, J.J. J. Econ. Entomol. 1982, 75, 637-641.
69. Ahmad, N.; Walgenbach, D.D.; Sutter, G.R. Bull. Environ. Contam. Toxicol. 1979, 23, 572-574.
70. Koeppe, M.K.; Lichtenstein, E.P. J. Agric. Food Chem. 1982, 30, 116-121.
71. Ballee, D.L., D.E. Stallard and J.A. Ignatoski; company document #392-3AS80-0052-005, Department of Safety Assessment, Diamond Shamrock Corp., Painesville, Ohio, 1981.
72. Burlinson, N.E.; Lee, L.A.; Rosenblatt, D.H. Environ. Sci. Technol. 1982, 16, 627-632.
73. Chiou, C.T.; Peters, L.J.; Freed, V.H. Science 1979, 206, 831-832.
74. Wilson, J.T.; Enfield, C.G.; Dunlap, W.J.; Cosby, R.L.; Foster, D.A.; Baskin, L.B. J. Environ. Qual. 1981, 10, 501-506.
75. Wolf, J.M. M.S. Thesis, University of California, Davis, 1967.
76. Castro, C.E.; Belser, N.O. Environ. Sci. Technol. 1968, 2, 779-783.
77. Cohen, S.Z. Summary Report - DBCP in Ground Water in the Southeast, Hazard Eval. Div. TS-769c, EPA, Wash. D.C. 20460, 8/12/81, 15 pages.
78. Carter, G.E.; Riley, M.B. Pestic. Monit. J. 1981, 15, 139-142.
79. Mink, J.F. "DBCP and EDB in Soil and Water at Kunia, Oahu, Hawaii", Feb. 1, 1981, under contract to Del Monte Corp., Honolulu, HI.
80. Zalkin, F.; Wilkerson, M.; Oshima, R.J.; "Pesticide Movement to Groundwater, Volume II. Pesticide Contamination in the Soil Profile at DBCP, EDB, Simazine and Carbofuran Application Sites" California Department of Food and Agriculture, 1983, Sacramento, California 95814; in review.
81. "Draft Trial Assessment for Test Rule Support: 1,2-Dichloropropane," U.S. Environmental Protection Agency, Contract Number 68-01-6150, 1982.
82. CRC Handbook of Chemistry and Physics, 57th ed., Weast, R.C., Ed., 1976-1977, p. D-193, CRC Press, Inc., Cleveland, Ohio.
83. Goring, C.L. An. Rev. Phytopathol. 1967, 5, 285-318.

84. McKenry, M.V.; Thomason, I.J. Hilgardia 1974, 42, 11.
85. "Census of Agriculture," vol. 4, Bureau of the Census, Dept. of Commerce, Washington, D.C., 1974.
86. "Water-Related Environmental Fate of 129 Priority Pollutants," vol. II, U.S. Environmental Protection Agency, EPA 440/4-79-029b, 1979.
87. Li, M.; Ali, S.; Burau, R.; Crosby, D.; Hatsfield, J.; Hsieh, D.; Kilgore, W.; Painter, W.; Seiber, J. in "A Systems Approach to Controlling Pesticides in the San Joaquin Valley"; Ecosystems Studies of National Science Foundation, University of California, Davis, CA, 1979.
88. Roberts, T.R.; Stoydin, G. Pestic. Sci. 1976, 7, 325-335.
89. Cohen, D.B.; Gilmore, D.; Fischer, C.; Bowes, G.W. "1,2-Dichloropropane (1,2-D) 1,3-Dichloropropene (1,3-D)", 1983, California State Water Resources Control Board, Sacramento, California.
90. Frissel, M.J.; Bolt, G.H. Soil Sci. 1962, 94, 284-291.
91. Swann, R.L.; McCall, P.J.; Unger, S.M. Am. Chem. Soc. Nat. Mtg. Pestic. Chem. Div. Abstracts 1979, Hawaii.
92. Wallnofer, P.R.; Ziegler, W.; Engelhardt, G.; Rothmeier, H. Chemosphere 1978, 12, 967-972.
93. Moran, D., Suffolk County Department of Health, Hauppauge, New York 11788, personal communication, March 1983.
94. Hamaker, J.W.; Kerlinger, H.O. in "Vapor Pressure of Pesticides in Pesticidal Formulations Research Physical and Colloidal Chemical Aspects"; Wallenburg, J.W., Ed.; Adv. Chem. Ser. 86, American Chemical Society: Washington, D.C., 1969.
95. Mingelgrin, U.; Gerstl, Z. J. Environ. Qual. 1983, 12, 1-11.
96. Hamaker, J.W.; Thompson, J.M., in "Organic Chemicals in the Soil Environment"; Goring, C.A.I.; Hamaker, J.W., Eds.; Marcel Dekker, Inc., New York, 1972; Vol. 1, Chapter 2.
97. Jungclauss, G.; Blair, R.; Cohen, S.Z., work in progress, Midwest Research Institute and EPA.
98. "Ethylene Dibromide (EDB) - Position Document 4," U.S. Environmental Protection Agency, Office of Pesticide Programs, 1983.
99. Weaver, D.J.; Sava, R.J.; Zalkin, F.; Oshima, R.J.; "Pesticide Movement to Ground Water, Volume I: Survey of Ground Water Basins for DBCP, EDB, Simazine and Carbofuran," California Department of Food and Agriculture, 1983, Sacramento, California 95814.
100. Senn, L.H. , College of Agricultural Sciences, Clemson University, 11/2/83, personnal communication.

101. Jovanovich, A.P.; Cohen, S.Z. "Monitoring Groundwater in Georgia for Ethylene Dibromide (EDB) - A Preliminary Reconnaissance in Seminole County, Georgia"; in review, TS-769c. Hazard Evaluation Division, OPP/EPA, Washington, D.C. 20460.
102. Harvey, J. Jr.; Han, J.C.-Y. J. Agric. Food Chem. 1978, 26, 536-541.
103. Leistra, M.; Bromilow, R.H.; Boesten, J.J.T.I. Pestic. Sci. 1980, 11, 379-388.
104. Jordan, L.S. Residue Rev. 1970, 32, vii-xiii.
105. Helling, C.S. Soil Sci. Soc. Am. Proc. 1971, 35, 743-748.
106. Khan, S.U.; Marriage, P.B. Weed Sci. 1979, 27, 238-241.
107. Best, J.A.; Weber, J.B. Weed Sci. 1974, 22, 364-373.
108. Slack, C.H.; Blevins, R.L.; Rieck, C.E. Weed Sci. 1978, 26, 145-148.
109. Enfield, C.G.; Carsel, R.F.; Cohen, S.Z.; Phan, T.; Walters, D.M. Ground Water 1983, 20, 711-722.
110. Carsel, R.F.; Smith, C.N.; Lorber, M.N. "Pesticide Root Zone Model: Transient Hydraulic Model," EPA/ORD, Athens, GA, in review.
111. Ruiz, C.; Phan, T.; Wagner, J.; Enfield, C.; Kent, D. "Computer Models for Two Dimensional Subterranean Flow and Pollutant Transport", in review.
112. Karickhoff, S.W., unpublished data.
113. For example, Horowitz (Revue Roumaine de Chimie, 1978, 23, 603-611) performed Huckel molecular orbital calculations on a substituted, symmetrical triazine which demonstrated the alternating charge system in the ring.
114. Brady, N.C. in "The Nature and Properties of Soils", 8th edition; Macmillan Publishing Co., Inc., New York, 1974; Chap. 4.
115. Kadoum, A.; Mock, D.E. J. Agric. Food Chem. 1978, 26, 45-50.
116. Spalding, R.F.; Exner, M.E.; Sullivan, J.J.; Lyon, P.A. J. Environ. Qual. 1979, 8, 374-383.
117. Brooks, G.T. J. Environ. Sci. Health 1980, B15, 755-793.
118. Crosby, D.G. in "The Photodecomposition of Pesticides in Water"; Gould, R.F., Ed.; Adv. Chem. Ser. No. 111, American Chemical Society: Washington, D.C.; pp. 173-188.
119. "Pesticide Assessment Guidelines Subdivision N, Chemistry: Environmental Fate," EPA-540/9-82-021, U.S. Environmental Protection Agency, 1982.
120. Skurlatov, Y.I.; Zepp, R.G.; Baughman, G.L. J. Agric. Food Chem. 1983, 31, 10651071.
121. Khan, S.U. J. Environ. Sci. Health 1980, B15, 1071-1074.

RECEIVED February 13, 1984

Transfer of Degradative Capabilities

Bacillus megaterium to *Bacillus subtilis* by Plasmid Transfer Techniques

JOHN F. QUENSEN, III, and FUMIO MATSUMURA

Pesticide Research Center, Michigan State University, East Lansing, MI 48824-1311

Degradative plasmids are extrachromosomal genetic elements that code for metabolic pathways in bacteria. Plasmids are believed to be partly responsible for the metabolic diversity of Pseudomonas spp. We have found that Bacillus megaterium also has versatile degradative capabilities, at least partially degrading such substrates as DDT, parathion, heptachlor, naphthalene, polychlorinated biphenyls (PCB's), and 2,3,7,8-tetrachlorodibenzo-p-dioxin (TCDD). By transformation of Bacillus subtilis with B. megaterium plasmid enriched DNA preparations we have implicated these plasmids in coding for some of B. megaterium's degradative capabilities. The potential uses of degradative plasmids in engineering bacterial strains better suited for the treatment of toxic wastes and environmental decontamination will be discussed.

It is widely recognized that the persistence of environmental contaminants in soils and sediments is related in large part to their resistance to microbial degradation. The more resistant to microbial attack, the longer such chemicals tend to persist in the environment (1). Thus there has been considerable interest in selecting and identifying microorganisms capable of attacking persistent compounds. Research in this area not only provides information on the possible ultimate fate of such compounds in the environment, but also identifies microorganisms that may be used in waste treatment processes. Research on microbial degradation has also revealed that some bacterial plasmids code for the degradation of xenobiotics. Bacterial plasmids are extrachromosomal, small, autonomously replicating covalently closed cyclic DNA molecules

0097–6156/84/0259–0327$06.00/0

(2) that typically code for some special or unknown functions. Known functions include the fertility or sex factor in E. coli, antibiotic resistance, the production of bacteriocins such as colicins and megacins, and degradative pathways. Plasmids that code for degradative pathways are termed degradative or catabolic plasmids. Several reviews on degradative plasmids have been published (3-6). The discovery of degradative plasmids has sparked interest in using such plasmids to genetically engineer new bacterial strains that are more capable of degrading certain substrates of interest (7-9).

Most of the degradative plasmids discovered so far have been found in Pseudomonas species, and it is believed that plasmids are in part responsible for the versatility of Pseudomonas in attacking a wide variety of carbon sources (10). Some of the degradative plasmids occurring in Pseudomonas are listed in Table I. These plasmids confer on their host the ability to degrade a specific compound, and most were named, originally, for that compound. For example, TOL codes for toluene degradation while CAM codes for camphor degradation.

Several characteristics of typical degradative plasmids may be given. They tend to be among the larger (60 - 140 megadaltons and more) of the known plasmids (6). They may code for the partial or complete degradation of a hydrocarbon (4). Such compounds may be aliphatic (e.g., octane) (11), aromatic (naphthalene) (12-13), or chlorinated hydrocarbons (parachlorobiphenyl) (14-15). While we are primarily interested in plasmids that code for the degradation of exotic compounds, it should be remembered that some plasmids code for the degradation of such innocuous substances as sugars (16). Another characteristic of plasmids is that the enzymes involved in a plasmid coded degradative pathway are inducible by the primary substrate. All of the genes involved are found in one or more operons activated by the primary substrate of the degradative pathway (4).

While most of the known degradative plasmids occur in Pseudomonas, the existence of degradative plasmids is probably much more widespread. At first plasmids were recognized only when they conferred a property of a priori interest that could be transferred independently of known chromosomal markers. This put strict limitations on the number and types of plasmids that could be discovered. More recently the techniques of equilibrium density centrifugation in a cesium chloride gradient and gel elctrophoresis have revealed that plasmids are ubiquitous (17-18). These techniques detect plasmids regardless of their function (hence the term cryptic plasmids for those of unknown function) and it is likely that some of them have degradative functions. As an example outside of the genus Pseudomonas, the ability of a strain of Klebsiella pneumonia to use parachlorobiphenyl as a sole carbon source has been attributed to a plasmid (15).

Plasmids and Genetic Engineering. There are several advantages in using plasmids to create pollutant degrading microorganisms over other genetic engineering techniques. For one, degradative pathways are complex. Several enzymes are required, along with cofactors and regulatory mechanisms. Degradative plasmids provide all of this in one relatively

Table I.   Some Typical Degradative Plasmids (4)

Plasmid	Degradative Pathway	Transmissibility[a]	Size (Megadaltons)
CAM	Camphor	+	> 100[b]
OCT	n,Octane	–	> 100[b]
SAL	Salicylate	+	55,48,42
NAH	Naphthalene	+	46
TOL	Xylene, toluene	+	76
XYL-K	Xylene, toluene	+	90
NIC	Nicotine, nicotinate	+	N.D.[c]
pJP1	2,4-Dichlorophenoxy-acetic acid	+	58
pAC21	p-chlorobiphenyl	+	65

[a] + indicates plasmid is transmissible
[b] exact size unknown
[c] N.D. not determined

easy to manage package (9). As mentioned above, the required genes exist in one or more operons activated by the primary substrate. Many plasmids are transmissible, meaning they may be transmitted between strains or even between species simply by culturing donor and recipient cells together. When this is not feasible, techniques exist for isolating plasmids and transferring them to recipient cells. A second, at least theoretical, advantage in using plasmids to engineer pollutant degrading microorganisms is that they may be used to genetically alter microorganisms already present in a polluted environment so that they gain the ability to better degrade a pollutant without losing their ability to survive field conditions. Pollutant degrading microorganisms developed by the more traditional procedure of selection in the laboratory often fail to become established at the polluted site because such strains lack the ability to survive field conditions and to compete with other existing microorganisms.

At the same time it must be recognized that there are still problems to overcome in using plasmids for genetic engineering purposes. Introduced DNA may be recognized by the new host cell as foreign and destroyed or modified in such a way that it is non-functional (19). Also, certain plasmids are incompatible and cannot coexist in the same cell. Presumably this occurs when the plasmids are closely related and under stringent copy number control (20). Barring these obstacles, there may remain poorly understood problems leading to the lack of expression of the plasmid genes in the new host. Attempts to transfer functional degradative plasmids from Pseudomonas species, which have been most frequently studied by scientists in this field, to other microbial species (12-13) usually have not been successful. In 1978 Chakrabarty et al. (21) showed that it is possible to transfer the TOL plasmid, which carries genes encoding enzymes to degrade xylenes, toluene, and trimethylbenzene derivatives, from Pseudomonas putida to E. coli. However, in E. coli the plasmid did not express degradative functions for toluene or salicylate. The fact that the TOL plasmid expresses such a function when it is transferred back to P. putida after cloning in E. coli indicates that expression of degradative DNA's may be controlled by the chromosomal DNA. Jacoby et al. (22) transferred the TOL plasmid from P. putida and P. aeruginosa to E. coli, but they also found that its functional expression was either very low or incomplete. These workers considered that some of the metabolic products from xylene and toluene are toxic to E. coli.

Despite these obstacles, the use of plasmids to create new strains of microorganisms with special degradative capabilities has already been demonstrated with some success. For example, by introducing plasmids coding for the degradation of several classes of compounds into the same microorganism Friello et al. (8) developed a new strain of Pseudomonas that can degrade a variety of the components of crude oil. The authors felt this approach was superior to using a mixed culture of microorganisms since it avoids the problem of temporal changes in the relative abundance of the different strains. A second example involves chlorobenzoate metabolism. Pseudomonas B13 can metabolize 3-chlorobenzoate, but not 4-chlorobenzoate because of the high substrate specificity of its 3-chlorobenzoate oxidase (23). By introducing a plasmid (TOL) coding for a benzoate 1,2-dioxygenase of broader substrate specificity into Pseudomonas B13, Knackmuss and his colleagues were able to derive a strain capable of metabolizing not only 3-chlorobenzoate, but also 4-chlorobenzoate and 3,5-dichlorobenzoate (23). In a similar experiment, Chatterjee and Chakrabarty (24) obtained a new plasmid, apparently coding for 4-chlorobenzoate degradation, from recombination between the TOL plasmid and a 3-chlorbenzoate degradative plasmid (pAC25) during selection in a chemostat. This method of combining strains with different degradative plasmids followed by selection in a chemostat has been termed "plasmid assisted molecular breeding" and has also led to the development of a 2,4,5-T degrading strain of Pseudomonas cepacia (25).

Work in the area of genetically engineering new strains to degrade specific hydrocarbons would be greatly facilitated if a system which freely allows the transfer and expression of genetic information in the recipient organism was known. We believe our research has the potential for identifying such a system as well as identifying new degradative plasmids from a gram positive microorganism (Bacillus megaterium) that may be used in future genetic engineering work.

Bacillus megaterium. We originally became interested in determining B. megaterium's ability to degrade a variety of xenobiotics because it possesses a cytochrome P-450 oxidative system (26-27) and degrades a wide variety of naturally occurring substrates including steroids, long chained fatty acids, amides, and alcohols (26, 28). We have shown that B. megaterium at least partially degrades DDT, parathion, heptachlor, chlordane (29), and 2,3,7,8-tetrachlorodibenzo-p-dioxin (30). In addition to its wide spectrum degradative ability, B. megaterium possesses many copies of a number of different plasmid-like DNA segments (31-33). We therefore wished to determine if this degradative ability is plasmid coded and whether it can be transferred to other bacteria.

Chracteristics of B. megaterium's plasmid system have been summarized by Carlton (31). The plasmids are typical in that they band as covalently closed circular DNA in ethidium bromide-cesium chloride gradients and they are resistant to irreversible alkaline denaturation (17). However, B. megaterium plasmids are atypical in that they exist in approximately 10 size classes and as many copies per cell (32). In fact, for the smaller plasmids there are hundreds of copies per cell so that plasmid DNA may represent up to 40% of the total extractable DNA (31). This is unusual since for most plasmids there are usually no more than a few copies per cell. Also, hybridization studies suggest that there is extensive homology between three B. megaterium plasmids of different sizes and between these plasmids and the chromosomal DNA (31,33). Carlton (31) concludes that the most likely explanation of the origin of B. megaterium plasmids is that they are molecular hybrids between one or more plasmid elements and various portions of the chromosomal DNA.

The only known function for any of the B. megaterium plasmids is megacin A production by the 30.9 megadalton plasmid (34). However, if Carlton's hypothesis on the origin of these plasmids is true, then we may expect many functions coded for by the chromosomal DNA to be duplicated by the plasmids. This should enhance our chances of being able to transform other bacterial species to degrade xenobiotics using B. megaterium plasmids.

Materials and Methods

Strains. Three derivative strains of Bacillus metagerium (ATCC 13368) were selected for resistance to dibenzofuran, parachlorobiphenyl, and naphthalene by plating the parent strain on nutrient agar plus 0.1% of each compound. These strains served as the DNA donors. While B. megaterium partially degrades each of these compounds, we were not successful in selecting a strain of B. megaterium to grow on any one as a sole carbon source. A streptomycin resistant derivative of Bacillus subtilis Marburg 168 RM 125 arg 15 leu A8 r_M^- m_M^- (35) was kindly supplied by Dr. Teruhiko Beppu (University of Tokyo) and Dr. Kiyoshi Miwa (Central Research Laboratories, Aijinomoto Co., Inc.) and was used as the DNA recipient.

Isolation of Plasmid Enriched DNA. The three strains of B. megaterium were cultured in yeast–soy broth (4% Bacto–yeast extract and 1.6% Bacto–soytone from Difco Laboratories, Detroit, pH 7.2). Plasmid DNA from each strain was isolated and purified by cesium chloride–ethidium bromide equilibrium density centrifugation following the procedure of Carlton and Brown (36). Subsequent gel electrophoresis (0.5% agarose) revealed the presence of some chromosomal DNA contamination.

Transformation Procedure. A streptomycin resistant strain of B. subtilis Marburg 168 was used as the recipient strain. The transformation procedure was the protoplast–polyethylene glycol (PEG) method described by Chang and Cohen (37). Basically, donor DNA is added to a suspension of B. subtilis protoplasts in the presence of PEG. Three transformation experiments were performed using DNA preparations from each of the three B. megaterium strains selected to grow on nutrient agar plus 0.1% of dibenzofuran, parachlorobiphenyl, or naphthalene. Following the regeneration of protoplasts on DM3 medium containing streptomycin (100 µg/ml) transformants were selected for their ability to grow more rapidly than the original B. subtilis strain on the corresponding selective medium. Because even B. megaterium could not use any of these substrates as a sole carbon source we were not able to select directly for degradative capability. However, we did expect some correlation between resistance and degradative capability. We therefore selected for resistance and screened the resistant transformants for degradative ability. Preliminary tests demonstrated that B. megaterium colonies appeared in less than 24 hours after innoculating the selective plates while B. subtilis required longer (generally 48 hours) to show growth. B. subtilis protoplasts regenerated without the addition of B. megaterium DNA also failed to grow on these selective media in less than 24 hours. Transformants selected in this manner were tested for the presence of B. subtilis chromosomal markers (auxotrophic for arginine and leucine).

Resistance Tests.   Several transformants were tested for their ability to grow on all three selective media.   These transformants were streaked onto nutrient agar plates containing 0.1% naphthalene, dibenzofuran, or parachlorobiphenyl.   The plates were incubated for 18 hours at 25°C before scoring.

Screening Transformants.   The naphthalene resistant B. subtilis transformants were screened for their ability to degrade naphthalene according to the method of Ensley et al. (38).   Cells were cultured in nutrient broth supplemented with naphthalene (1 mg/ml), filtered through glass wool to remove naphthalene crystals, harvested by centrifugation, washed twice with 0.05 M potassium phosphate buffer, pH 7.2, and suspended at a concentration of 40 mg (wet cell weight) per milliliter of buffer.   The reaction system consisted of 5 ml of cell suspension, 0.05 μCi ^{14}C-naphthalene plus 100 μg cold naphthalene in 100 μl ethyl acetate in tightly capped culture tubes to prevent volatilization of the naphthalene.   Naphthalene degradation was detected by the accumulation of nonvolatile metabolites of ^{14}C-naphthalene.   After ten hours of incubation at room temperature on a reciprocal shaker, 20 μl samples of the reaction mixtures were spotted on pieces of glass silica gel thin layer chromatography plates (1.5 X 1.0 cm) and the remaining naphthalene was removed under a stream of nitrogen for 15 minutes.   The ^{14}C activity remaining on the squares was determined by liquid scintillation counting.   This remaining activity respresented nonvolatile metabolites of ^{14}C-naphthalene (38).   Controls were done similarly using autoclaved cells.   The controls take into account all possible fates of the naphthalene other than metabolism.   These possibilities include binding to cells or debris, adsorption to the flasks or plugs, sublimation, autocatalysis, hydrolysis, etc.   Differences between control and experimental units must therefore be due to metabolism.

Assays of Naphthalene Degradation.   Washed cell experiments were used to compare the naphthalene degrading abilities of B. megaterium (both unselected and selected with naphthalene), B. subtilis, and transformants which showed degradative ability in the screening tests.   Washed cells were prepared as described above and suspended at a concentration of 25 mg (Experiment 1) or 6 mg (Experiment 2) of cells per milliliter of buffer.   Autoclaved cell suspensions were used as controls and account for all possible fates of the naphthalene other than metabolism.

Incubation of the reaction mixture was done in screw capped tubes to eliminate losses of naphthalene due to volatilization.   The reaction system for Experiment 1 consisted of 10 ml of cell suspension, 0.05 μCi of ^{14}C-naphthalene, and 100 μg of cold naphthalene in 200 μl of ethyl acetate.   The reaction mixture for Experiment 2 consisted of 5 ml of cell suspension and 0.8 μCi of ^{14}C-naphthalene in 50 μl of ethyl acetate.   The mixtures were incubated at room temperature for 10 hours with gentle shaking.   The reaction mixtures were then extracted three times with equal volumes of ethyl acetate (Experiment 1) or hexane:acetone (4:1, v:v) (Experiment 2).   A statistically significant increase in non-extractable ^{14}C activity (primarily polar, water soluble metabolite) for live cells over control dead cells was used as the criterion for degradation. These ^{14}C activities were determined by liquid scintillation counting. The statistical comparisons between the live and dead cell (control) treatments were made using a one tailed t test, correcting for unequal variances when indicated. Differences were declared significant if the probability of a larger | t | was less than 0.05 (i.e., P = (1 -α) > 0.95).

TLC Analysis. For experiment 2, thin layer chromatographic analysis was carried out by reducing solvent volumes to less than 250 µl under a nitrogen stream, spotting on activated silica gel plates, and developing in chloroform:acetone (20:1, v:v). Autoradiographs of these plates were prepared by exposing them to X-ray film for 3 or 4 days at -20°C.

Results

Bacillus subtilis transformants were identified by their resistance to naphthalene, parachlorobiphenyl, or dibenzofuran. Controls (B. subtilis protoplasts regenerated without the addition of B. megaterium plasmid DNA) exhibited no increased reistance to these three compounds. That each transformant was B. subtilis and not a contaminant was verified by testing for the B. subtilis chromosomal markers. Only streptomycin resistant strains that showed arginine and leucine dependent growth were used in subsequent experiments.

Hydrocarbon Resistance. The importance of the source of the DNA (B. megaterium selected on naphthalene, parachlorobiphenyl, or dibenzofuran) on the transformants' abilities to grow on media containing each hydrocarbon was determined. DNA from selected strains of B. megaterium was transferred to B. subtilis in three separate transformation experiments. The growth response of the resulting transformants on nutrient agar with or without 0.1% of each hydrocarbon was studied at 24°C. The data (Table II) clearly show that in this transformation system hydrocarbon resisting or degrading abilities which are controlled by the B. megaterium DNA are transferable and are functionally expressed in the recipient system, and also that the substrate specificity acquired by selection prior to transformation is still retained in the recipient strains after transformation. It is interesting to note that the transformants with DNA from B. megaterium selected on dibenzofuran showed some tendency to grow also on the medium containing naphthalene and parachlorobiphenyl in this experiment.

Naphthalene Degradation. Twelve transformants showing faster growth on the selective medium plates were screened for their ability to degrade naphthalene according to the method of Ensley et al. (38). Three of these were selected for further study based on their production of nonvolatile ^{14}C-naphthalene metabolites. The naphthalene degradation abilities of B. megaterium, B. subtilis, and the transformants Nah 1, Nah 25, and Nah 28 were compared in washed cell experiments using ^{14}C-naphthalene.

Both disappearance of solvent extractable radiocarbon and accumulation of water soluble, nonextractable radiocarbon were taken as

Table II.   Growth Response after 18 Hours of B. subtilis Transformants
with B. megaterium DNA

Microorganisms	Nutrient Agar Containing		
	Nah	Dbf	pCB
Original microorganisms			
B. megaterium	+	+	+
B. subtilis	–	–	–
B. megaterium selected with:			
Naphthalene (Nah)	++	+	+
Dibenzofuran (Dbf)	+	++	+
p–chlorobiphenyl (pCB)	+	+	++
Transformants with:			
Naphthalene selected DNA			
Nah 27	+	–	–
Nah 28	+	–	–
Dibenzoburan selected DNA			
Dbf 3	s	+	s
Dbf 5	s	+	s
p–chlorobiphenyl selected DNA			
pCB 15	–	–	+
pCB 16	–	–	+

"++" indicates vigorous growth and/or growth started earlier than others
"+" indicates moderate growth
"s" indicates slight or questionable growth
"–" indicates no growth

indications of naphthalene degradation. In general, a decrease in solvent extractable radiocarbon was paralleled by an increase in nonextractable, water soluble radiocarbon.   However, the greater variation associated with the solvent extractable radiocarbon counts, most likely due to volatilization of naphthalene, made statistically significant differences more difficult to demonstrate. For this reason we considered a difference between live cells and the controls in the nonextractable ^{14}C activity (Table III) as the more reliable indication of naphthalene metabolism. Column P_{14} in Table III gives the probabilities that the differences in aqueous ^{14}C activities recovered between live and dead cells are statistically significant. In Experiment 1 (three replicates) there was no evidence of naphthalene degradation by B. subtilis or unselected B. megaterium.   The naphthalene selected B.   megaterium did show a significant (P=0.98) increase in nonextractable ^{14}C activity, and the transformant Nah 28 showed a slight but statistically nonsignficant (P=0.93) increase. However, in the second experiment (two replicates) all three transformants tested showed statistically significant increases in water soluble metabolites, with Nah 1 and Nah 28 giving higher yields. In this experiment B. subtilis also showed an increase in water soluble metabolites, although it was nonsignificant (P=0.94) and of less magnitude than for transformants Nah 1 and Nah 28.

The thin layer chromatographic analysis (Figure 1) revealed four solvent extractable metabolites present in the cases of B. megaterium, Nah 28, and three each in the cases of B. subtilis and Nah 1. The metabolite B (R_f=0.37) was unique to B. megaterium and the transformants Nah 1 and Nah 28 and occurred in greater quantity as judged by the comparative densities of the spots in the autoradiographs. As little as 0.4 nCi (0.05% of the input ^{14}C activity) can be detected in the autoradiographs.

Discussion

That the traits of B. megaterium can be transferred to B. subtilis by plasmid transfer techniques has been established.   These traits are increased resistance to naphthalene, parachlorbiphenyl, and dibenzofuran, and increased ability to degrade ^{14}C-naphthalene. The transformants' increased degradative abilities were demonstrated by the accumulation of greater quantitites of water soluble metabolite and the presence of a unique solvent soluble metabolite.

The transformation techniques employed were designed to transfer plasmid DNA from B. megaterium to B. subtilis and the results suggest that these plasmids are involved in coding for B. megaterium's degradative abilities. Bacillus megaterium contains many plasmids (at least nine according to Carlton and Brown (36), six of which comprise the bulk of closed circular plasmid DNA. Their sizes are listed as 4, 6.2, 15.9, 30.9, 47, and 60 megadaltons. If any of these plasmids are involved in determining B. megaterium's degradative abilities, then they may prove important in engineering microorganisms to degrade organic pollutants. However, at this time we have not been able to demonstrate the presence of B. megaterium plasmids in the transformants, nor can we rule out the

Table III. Recovered ^{14}C Activity (Average Disintegrations per Minute $\pm$ Standard Deviation) for the Naphthalene Degradation Experiments

Strain	^{14}C-Activity Recovered				P
	Solvent Phase		Aqueous Phase		
	Dead Cells	Live Cells	Dead Cells	Live Cells	
Experiment I					
B. megaterium, not selected	38,592 ± 1,242	37,041 ± 2,329	1,042 ± 264	1,013 ± 81	0.001
B. megaterium naphthalene selected	36,048*	29,902 ± 1,062	1,595 ± 302	3,279 ± 1,317	0.98
B. subtilis	38,221 ± 2,698	33,014 ± 5,883	2,665 ± 879	1,888 ± 363	0.11
Nah 28	38,805 ± 516	38,760 ± 934	910 ± 396	1,394 ± 232	0.93
Experiment II					
Nah 1	593,459 ± 33,901	518,404 ± 41,218	399 ± 282	3,648 ± 183	0.99
Nah 25	547,051 ± 1,927	530,747 ± 7,653	647 ± 42	1,203 ± 121	0.99
Nah 28	564,348 ± 18,776	596,526 ± 12,161	648 ± 12	3,540 ± 843	0.98
B. subtilis	575,405 ± 6,247	551,836 ± 35,272	670 ± 188	1,958 ± 672	0.94

*Only one observation.

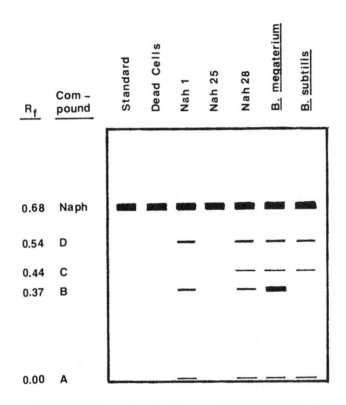

Figure 1.   Autoradiograph of thin layer chromatography plate showing
            solvent extractable naphthalene metabolites (A, B, C, D)
            produced by transformants Nah 1, Nah 25, Nah 28, B.
            megaterium, and B. subtilis.

possibility of chromosomal contamination of our DNA preparations and the resulting possibility that our transformants contain B. megaterium DNA. This may be a moot point, though, as DNA hybridization studies suggest considerable sequence homology between the total plasmid DNA and chromosomal DNA (31). Therefore, there is a possibility that B. megaterium plasmids harbor genetic information that also resides on the chromosome.

Degradation of hydrocarbons requires quite complex enzyme systems (13, 39, 40). Therefore, to be functional in the recipient cell, first many associated and unassociated genes involved in degradation must be transferred or already present in the recipient. Second, inhibitory factors such as inhibitory genetic regulatory systems must be absent in the recipient cells. Third, sufficient amounts of necessary cofactors (e.g., NADPH, FMN, FAD, etc.) must be made available. And fourth, neither the hydrocarbons nor their degradation products can be toxic to the recipient cells. It appears from our experimental results that the most readily transferrable genes are those conferring resistance to the hydrocarbons. Among such transformants a varying degree of degradation capabilities has been observed. Three of the twelve transformants produced nonvolatile ^{14}C-activity. Only two of these (Nah 1 and Nah 28) gave solvent soluble metabolites. Nah 25 also produced less nonextractable ^{14}C activity than the other two. This could be interpreted to mean that some transformants received most of the genetic information necessary for degradation, while others received only some portion.

While we did not test sufficient numbers of regenerated protoplasts to accurately determine the transformation frequency to aromatic hydrocarbon resistance, these frequencies were certainly high. Chang and Cohen (37) reported transformation frequencies of 10-80% using the same lysozyme-PEG method and similar quantities of plasmid DNA. The naphthalene transformants retained their degradative ability through at least six successive transfers.

As this research is still in progress we currently are unable to answer questions related to the extent of naphthalene, parachlorbiphenyl, or dibenzofuran degradation by the transformants, or whether the transformants are able to use any of these compounds as carbon or energy sources. While there are many other unanswered questions, it is clear from the current work that this transformation system may be used to explore many avenues of investigation relative to the microbial degradation of complex hydrocarbons. Obviously one of the future possibilities is to study the potential for developing such technology to degrade unwanted organic pollutants by using such transformation systems.

Acknowledgments

This work was supported in part by the Michigan Agricultural Experiment Station (Journal Article No. 11240) and the Michigan State Biomedical Program, Michigan State University, East Lansing, Michigan. S. Y. Oh performed the DNA extraction and transformation portions of the laboratory work while supported by the University of Agriculture Malaysia.

Literature Cited

1. Alexander, M. Science 1981, 211, 132-138.
2. Davis, D. B.; Dulbecco, R.; Eisen, H. N.; Ginsberg, H. S.; Wood, W. B. "Microbiology"; Harper and Row: New York, 1983.
3. Chakrabarty, A. M. Ann. Rev. Genet. 1976, 10, 7-30.
4. Farrell, R.; Chakrabarty, A. M. in "Plasmids of Medical, Environmental, and Commercial Importance"; Timmis, K. N.; Puhler, A., Ed.; Elsevier/North Holland Biomedical Press, 1979; 97-109.
5. Wheelis, M. L. Ann. Rev. Microbiol. 1975, 29, 505-524.
6. Williams, P. A. TIBS 1981, 6, 23-26.
7. Franklin, F. C. H.; Bagdasarian, M.; Timmis, K. N. in "Microbial Degradation of Xenobiotics and Recalcitrant Compounds"; Leisinger, T.; Hutter, R.; Cook, A. M.; Nuesch, J., Ed.; FEMS SYMPOSIUM NO. 12, Academic Press: London, 1981; pp. 109-130.
8. Friello, D. A.; Mylroie, J. R.; Gibson, D. T.; Rogers, J. E.; Chakrabarty, A. M. J. Bacteriol. 1976, 127, 1217-1224.
9. Williams, P. A. in "Microbial Degradation of Xenobiotics and Recalcitrant Compounds"; Leisinger, T.; Hutter, R.; Cook, A. M.; Nuesch, J., Ed.; FEMS SYMPOSIUM NO. 12, Academic Press: London, 1981; pp. 97-107.
10. Gunsalus, J. C.; Hermann, M.; Toscano, W. A.; Katz, D.; Garg, G. K. in "Microbiology - 1974"; Schlessinger, D. Ed. American Society of Microbiology: Washington, D. C., 1974; pp. 206-212.
11. Chakrabarty, A. M.; Chou, G.; Gunsalus, J. C. PNAS 1973, 70, 1137-1140.
12. Dunn, N. W.; Gunsalus, I. C. J. Bacteriol. 1973, 114, 974-979.
13. Williams, P. A.; Murray, K. J. Bacteriol. 1974, 120, 416-423.
14. Fisher, P. R.; Appleton, J.; Pemberton, J. M. J. Bacteriol. 1978, 135, 798-804.
15. Kamp, P. F.; Chakrabarty, A. M. in "Plasmids of Medical, Environmental, and Commercial Importance"; Timmis, K. N.; Puhler, A., Eds.; Elsevier/North Holland Biomedical Press, 1979; pp. 275-285.
16. Guiso, N.; Ullman, A. J. Bacteriol. 1976, 127, 691-697.
17. Carlton, B. C.; Helinski, D. R. PNAS 1969, 64, 592-599.
18. Helinski, D. R.; Clewell, D. B. Ann. Rev. Biochem. 1971, 40, 899-942.
19. Boyer, H. W. Ann. Rev. Microbiol. 1971, 25, 153-176.
20. Meynell, G. C. "Bacterial Plasmids"; MacMillan Press, Ltd.: London, 1972.
21. Chakrabarty, A. M.; Friello, D. A.; Bopp, L. H. PNAS 1978, 75, 3109-3112.
22. Jacoby, G. A.; Rogers, J. E.; Jacob, A. E.; Hedges, R. W. Nature 1978, 274, 179-180.
23. Reineke, W. R.; Knackmuss, H. J. J. Bacteriol. 1980, 142, 467-473.

24.   Chatterjee, D. K.; Chakrabarty, A. M. in "Microbial Degradation of Xenobiotics and Recalcitrant Compounds"; Leisinger, T.; Hutter, R.; Cook, A. M.; Nuesch, J., Eds.; FEMS SYMPOSIUM NO. 12, Academic Press, Inc.: London, 1981; pp. 213-219.

25.   Kilbane, J. J.; Chatterjee, D. K.; Karns, J. S.; Kellogg, S. T.; Chakrabarty, A. M. Appl. Environ. Microbiol. 1982, 44, 72-78.

26.   Berg, A.; Carlstrom, K.; Gustafsson, J.; Sundberg, M. I. Biochem. Biophys. Res. Comm. 1975, 66, 1414-1423.

27.   Hare, R. S.; Fulco, A. J. Biochem. Biophys. Res. Comm. 1975, 65, 665-672.

28.   Miura, Y.; Fulco, A. J. Biochemica et Biophysica Acta 1975, 388, 305-317.

29.   Yoneyama, K.; Matsumura, F., Unpublished data.

30.   Quensen, J. F., III; Matsumura, F. Environm. Toxicol. Chem. 1983, 2, 261-268.

31.   Carlton, B. C. in "Microbiology"; Schlessinger, D., Ed.; American Society of Microbiology: Washington, D. C., 1976; pp. 394-405.

32.   Carlton, B. C.; Smith, M. P. W. J. Bacteriol. 1974, 117, 1201-1209.

33.   Henneberry, R. C.; Carlton, B. C. J. Bacteriol. 1973, 114, 625-631.

34.   Rostas, K.; Dobritsa, S. V.; Dobritsa, A. P.; Koncy, C.; Alfoldi, L. Mol. Gen. Genet. 1980, 180, 323-329.

35.   Uozumi, T.; Hoshino, T.; Miura, K.; Horinouchi, S.; Beppu, T.; Arima, K. Molec. Gen. Genet. 1977, 152, 65-69.

36.   Carlton, B. C.; Brown, B. J. Plasmid 1979, 2, 59-68.

37.   Chang, S.; Cohen, S. N. Molec. Gen. Genet. 1979, 168, 111-115.

38.   Ensley, B. D.; Gibson, D. T.; LaBorde, A. L. J. Bacteriol. 1982, 149, 948-954.

39.   Dagley, S. in "Degradation of Synthetic Organic Molecules in the Biosphere"; National Academy of Science: Washington, D. C., 1972; pp. 1-16.

40.   Gunsalus, I. C.; Marshall, U. P. CRC Crit. Rev. Microbiol. 1971, 1, 291-310.

RECEIVED April 24, 1984

Degradation of High Concentrations of a Phosphorothioic Ester by Hydrolase

R. HONEYCUTT, L. BALLANTINE, H. LeBARON, D. PAULSON, and V. SEIM—
CIBA-GEIGY Corporation, Greensboro, NC 27409

C. GANZ—EN-CAS Laboratories, Winston-Salem, NC 27103

G. MILAD—Biospherics, Inc., Rockville, MD 20852

Greenhouse soil was treated at 500, 1000, 2000 and 5000 ppm
with Diazinon 4E. Parathion hydrolase was added to the soil
to determine the efficacy of the enzyme to rapidly degrade
diazinon during a spill situation. The half-life of the diazinon in the 500 ppm treatment without enzyme present was 9.4
days while the half-life of diazinon in the 500 ppm treatment
with enzyme present was one hour. The half-lives of diazinon
in the 1000, 2000 and 5000 ppm treatments with enzyme present
were 1.2, 5.6 and 128 hours (5.3 days), respectively. These
data indicate that parathion hydrolase can be used effectively
to rapidly reduce large concentrations of diazinon in soil.
At diazinon concentrations above 2000 ppm the enzyme is less
effective. Parathion hydrolase is readily soluble in water,
is reasonably stable and can be easily handled in the field.
Further research is needed to evaluate the efficacy of parathion hydrolase to decontaminate diazinon under actual spill
conditions.

Chemical spills can be devastating to the environment as well
as to the pocketbook of those responsible for its cleanup. The
cost of cleanup of a single spill can approach $200,000. It
immediately occurs to one that a much easier and cheaper way
must exist to accomplish the cleanup of chemical spills. At
CIBA-GEIGY we have formed a task force to study alternate ways
to clean up spills.

One of the first projects the task force undertook was to
develop a simple inexpensive means to clean up diazinon spills
which may occur on land or in water.

0097–6156/84/0259–0343$06.00/0
© 1984 American Chemical Society

Chakrabarty has extensively reviewed the biodegradation of pesticides (1). Table I shows the results of several studies on the enzymatic activity of microbial cell-free extracts for pesticide degradation. Clearly, there is substantial evidence to suggest that enzymes might be used in the development of biotechnology for use in degradation of pesticides.

Diazinon (Figure 1) is widely used throughout the United States and other countries for control of various insects such as cutworms, grubs, ants, cockroaches and silverfish. In the past, diazinon has been degraded using acid or sodium hypochlorite degradation (2,3). Recently, enzymatic hydrolysis for decontamination of diazinon spills has been studied by Munnecke, et al. (4,5). Using an enzyme, parathion hydrolase, from a mixed culture of pseudomonas sp., Munnecke was able to show that parathion hydrolase would degrade the organophosphate parathion 2450 times faster than 0.1N NaoH at 40°C (4). Table II shows the comparative hydrolysis rates of several organophosphates using parathion hydrolase vs. a chemical hydrolysis method. Parathion hydrolase outperforms chemical hydrolysis methods for most organophosphates that Munnecke looked at. Subsequent studies showed that the enzyme would degrade 1000 ppm diazinon (25% EC formulation) in soil by 97% in 24 hours (5).

Most of the studies done by Munnecke were small scale laboratory studies. The efficacy of parathion hydrolase has not been tested under field conditions. It was the major objective of our study to determine the usefulness of parathion hydrolase for the decontamination of high concentrations of formulated diazinon in soil under greenhouse conditions. A secondary, but very important, objective was to determine if the enzyme could be handled in a practical fashion as would be done in the field and retain its ability to degrade diazinon.

METHODS

Materials: Parathion hydrolase was obtained from Doug Munnecke (4,5). The specific activity was measured by a method of Munnecke (4,5) and was found to be 0.1 μmole diazinon hydrolyzed/mg total protein/minute. Diazinon 4E was obtained from CIBA-GEIGY, Greensboro, NC.

Description of Greenhouse Study - In 1981, six rectangular flats (2' X 3' X 3") of Georgia loamy sand soil were prepared. The soil in each flat was then placed in a large Hobart mixer (80 quart bowl) and the appropriate amount of Diazinon 4E

Table I. Enzymatic Activity of Microbial Cell-Free Extracts
For Pesticide Degradation

Pesticide	Enzyme Class	Enzyme Activity (NMOL of Substrate Transformed/Min/ mg Protein)
Organaophospates		
Acephate	Esterase	52
Aspon	Esterase	110
Cyanophos	Esterase	58
Diazinon	Esterase	301,1200,A
Dursban	Esterase	600
DCVMP	Lyase	A
DEFP	Lyase	A
DFP	Lyase	A
EPN	Esterase	12
Fenitrothion	Esterase	217
Fensulfothion	Esterase	238
Methyl Parathion	Esterase	600,A
Monocrotophos	Esterase	133
Paraoxon	Esterase	13,3600,A
Parathion	Esterase	259,3000, 7000,A
Propetamphos	Esterase	50
Quinalphos	Esterase	1410
Triazophos	Esterase	4350
TEPP	Lyase	A
Phenylureas		
Carboxin	Acylamidase	252
Chlorbromuron	Acylamidase	11,15
Linuron	Acylamidase	18,20,130
Metabromuron	Acylamidase	16,18
Monalide	Acylamidase	29,238
Monolinuron	Acylamidase	15,20
Monuron	Acylamidase	4
Pyracarbolid	Acylamidase	35

Note: A = insufficient data to calculate enzyme activity.

Table II.   Parathion Hydrolase Studies. Enzymatic vs. Chemical
            Hydrolysis of Organophosphate Insecticides

Pesticide	Solution Concentration MM	Ratio Enzymatic Hydrolysis Rate/ Chemical Rate
Parathion	45	2450
Triazophos	110	1005
Paraoxon	50	525
EPN	2	11
Diazinon	72	143
Methyl Parathion	150	122
Dursban	8	40

(47.5% active ingredient) was slowly mixed into it over a period of 5-10 minutes. At this point, 2 liters of a solution of
parathion hydrolase in 10 mM trihydroxy-methyl-amino methane
buffer (TRIS) - 5 mM cobalt chloride, pH 8.5, were sprayed onto
the soil using a 2-gallon garden hand sprayer. The mixture was
then mixed for an additional 15-30 minutes to achieve homogene-
ity, at which time 300 grams of soil were taken for analysis.
These are referred to as 0.5 hr. samples. Subsequent soil
samples were taken at 2, 4, 24 and 48 hours, 7 days, and 3
weeks.
        Table III shows the description of soil flats for the
greenhouse study. Between sampling periods all flats were kept
in the greenhouse at 24-27°C and 50-70% relative humidity.

Table III.   Diazinon Concentrations in Georgia Greenhouse Soil
                            Flats

Soil Flat A - Enzyme Only
Soil Flat B - Diazinon 500 ppm - No Enzyme
Soil Flat C - Diazinon 500 ppm + Enzyme
Soil Flat D - Diazinon 1000 ppm + Enzyme
Soil Flat E - Diazinon 2000 ppm + Enzyme
Soil Flat F - Diazinon 5000 ppm + Enzyme

All flats kept at 24-27°C and 50-70% relative humidity.

Analytical Methods

Diazinon concentrations in soil were measured using extraction
- partition followed by gas chromatography. Figure 2 gives an
outline of this procedure. Fifty grams of soil was first
extracted with 200 ml acetone. Water (500 ml) was added to a
140 ml aliquot of the extract and the mixture partitioned with
35 ml hexane. The hexane was analyzed for diazinon using by
GLC and a flame photometric detector (Figure 3). Oxypyrimidine, the major product of diazinon degradation, (4 hydroxy-6-
methyl-2-isopropyl-pyrimidine) was determined in soil by extraction with methanol:water (80:20) followed by analysis by
HPLC using a reverse phase Whatman partisil ODS-3 column and a
50/50 mixture of methanol and $0.01M$ NaH_2PO_4 in the isocratic
mode (Figure 4).

Results

Solubility of Parathion Hydrolase During Soil Application

Parathion hydrolase is soluble in water and easily handled
during application to soil.

     For practical use on soil, it was found that large amounts
of the enzyme should be dissolved into a minimum amount of 10
mM TRIS - 5 mM cobalt chloride, pH 8.5, and the mixture blended
with a Waring blender at low speed for 1-2 minutes to achieve a
homogenous suspension. At this time more buffer can be added
gradually with constant agitation to finally achieve the desired dilute solution. This is not a complicated procedure.
Most workers will have ready access to the equipment needed to
carry out the procedure. Some settling of the enzyme will
occur over a period of time. However, the solution remains
homogeneous enough to pass through common garden spray equipment. This would make the enzyme quite practical for use in
ground application equipment.

Stability of Parathion Hydrolase

The stability of parathion hydrolase was determined for refri-
gerator storage and room temperature storage. The enzyme solution is stable under refrigeration (4). The enzyme also
remains stable at room temperature for long periods (4).

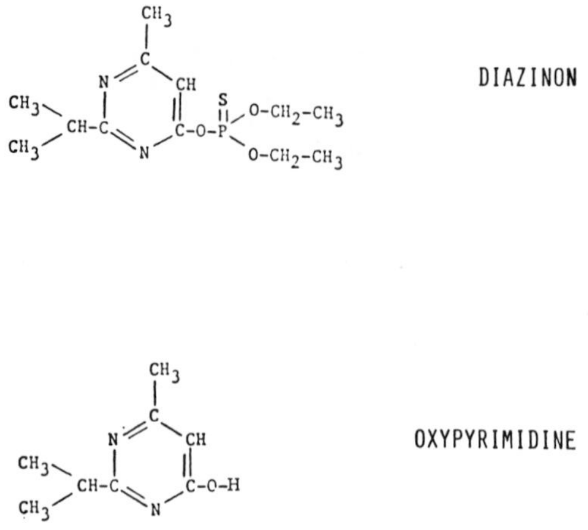

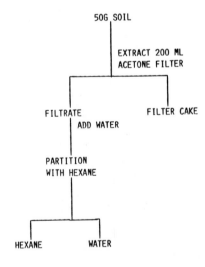

DIAZINON

OXYPYRIMIDINE

Figure 1.   Chemical Structures

Figure 2.   Analytical Methods to Determine Diazinon in Soil

COLUMN 6'X2MM GLASS PACKED WITH 10% DC-200 ON 80/100 GAS
CHROM Q

OVEN TEMPERATURE = 170°L

GAS FLOWS: CARRIER (HE) - 40 ML/MIN.
            HYDROGEN   - 50 ML/MIN.
            AIR   - 100 ML/MIN.

DETECTOR: FLAME PHOTOMETRIC (SULFUR MODE)

Figure 3.   GC Conditions for Analysis of Diazinon

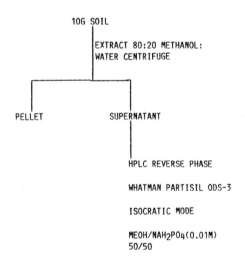

Figure 4.   HPLC Analysis of Oxypyrimidine in Soil

Degradation of Diazinon in Greenhouse Soil by Parathion Hydrolase

Parathion hydrolase degrades diazinon rapidly at high concentrations in soil. Table 4 shows a summary of the results of
the attempt to degrade high concentrations of diazinon with
parathion hydrolase. The first column shows that diazinon
degrades very slowly at 500 ppm with no enzyme present. The
subsequent columns show that parathion hydrolase rapidly and
effectively degrades high concentrations of diazinon. At 2000
ppm the half-life of diazinon, when enzyme is present, is about
4-5 hours. At 5000 ppm the enzyme is not as effective. This
may be due to the large amount of Diazinon 4E present which
could inhibit the enzyme.

Table IV. Efficacy of Parathion Hydrolase Decontamination of
           Diazinon 4E Georgia Sandy Loam Soil

Average ppm of Diazinon Remaining in Soil from Treatments[1,2]

Time After Application	500 (No Enzyme)	500	1000	2000	5000
			(With Enzyme Present)		
0.0 Hour	500[3]	500[3]	1000[3]	2000[3]	5000[3]
0.5 Hour	618+ 91	102+13	524+19	2081+209	6770+687
2.0 Hours	398+133	33+ 8	151+ 7	1281+339	4891+435
4.0 Hours	627+ 25	15+ 2	87+ 2	1317+ 13	6228+190
1 Day	547+ 26	3+ 0	4+ 1	40+ 0	4456+605
2 Days	634+ 12	2+ 0.7	4+ 0.7	7+ 0.7	3084+118
1 Week	501+ 9	<2	65+ 4	59+ 8	1228+ 49
3 Weeks	109+ 11	<1	<1	<1	397+ 0

[1]All values are averages of 2 replicate 0-3" soil core samples.
[2]Recoveries ranged from 67-129% (mean = 92.7%$\pm$14.1)
[3]Theoretical level of diazinon.

Figure 5 shows the degradation of diazinon in soil at 2000 ppm
by parathion hydrolase. It is easy to see that degradation is
very rapid and effective - reaching safe environmental levels
by 24 hours.

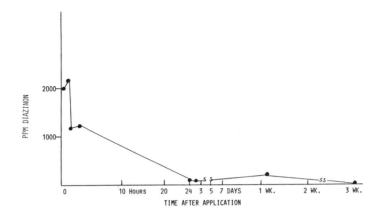

Figure 5. Degradation of Diazinon by Parathion Hydrolase

        One important aspect of any cleanup technique, is the type
of degradation products that are produced. These products must
be known in order to assess their potential environmental impact and toxicological hazards. One of the major degradation
products of diazinon, oxypyrimidine was measured in soil after
treatment with parathion hydrolase. Figure 6 shows that oxypyrimidine increases in soil as the diazinon is degraded by the
enzyme.

CONCLUSIONS AND SUMMARY

1. The enzyme parathion hydrolase is active enough to be used
    effectively to degrade high concentrations of diazinon in
    soil. The half-life of diazinon in soil treated at 2000
    ppm was 5.6 hours.
2. Oxypyrimidine is one of the degradation products after
    using parathion hydrolase.
3. The enzyme is soluble and stable, does not clog sprayers,
    and is easy and practical to use in the field.
4. Further studies are necessary to define the field para-
    meters which afford maximum activity of the enzyme (e.g.,
    soil type, amount of moisture, whether or not multiple
    applications of enzyme are needed).
5. Further studies are needed to determine availability and
    costs of producing large quantities of enzyme for field
    use.

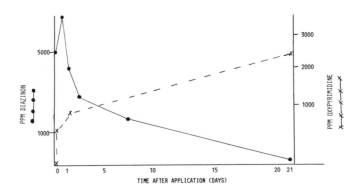

Figure 6.   Formation of Oxypyrimidine

Literature Cited

1. Chakrabarty, A. M. "Biodegradation and Detoxification of
    Environmental Pollutants." CRC Press, Inc. Boca Raton,
    Florida. 1982.
2. Dennis, W., E. Meier, W. Randall, A. Rosencrance,
    D. Rosenblatt. "Degradation of Diazinon by Sodium
    Hypochlorite, Chemistry and Tox." Env. Science & Tech.,
    13, 594 (1979).
3. Dennis, W., A. Rosencrance, W. Randall, E. Meier. "Acid
    Hydrolysis of Military Standard Formulation of Diazinon."
    J. of Env. Sci. & Health, Part B, Pesticides, Food Contam.
    & Ag Wastes, 1315(1), 47-60 (1980).
4. Munnecke, D. "Enzyme Hydrolysis of Organophosphate
    Insecticides, a Possible Disposal Method." J. App.
    Environ. Microbiol., Series 32, Issue 1, 7-13 (1976).
5. Munnecke, D. and S. Bank. "Enzymatic Hydrolysis of
    Diazinon in Soil." Dept. of Bot. & Micro., Univ. of
    Oklahoma, May 26, 1981.

RECEIVED March 12, 1984

INDEXES

Author Index

Subject Index

355

Production by Anne Riesberg
Indexing by Susan Robinson
Jacket design by Pamela Lewis

Elements typeset by Hot Type Ltd., Washington, D.C.
Printed and bound by Maple Press Co., York, Pa.

RECENT ACS BOOKS

"Stable Isotopes in Nutrition"
Edited by Judith R. Turnlund and Phyllis E. Johnson
ACS SYMPOSIUM SERIES 258; 240 pp.; ISBN 0-8412-0855-7

"Bioregulators: Chemistry and Uses"
Edited by Robert L. Ory and Falk R. Rittig
ACS SYMPOSIUM SERIES 257; 286 pp.; ISBN 0-8412-0853-0

"Polymeric Materials and Artificial Organs"
Edited by Charles G. Gebelein
ACS SYMPOSIUM SERIES 256; 208 pp.; ISBN 0-8412-0854-9

"Pesticide Synthesis Through Rational Approaches"
Edited by Philip S. Magee, Gustave K. Kohn, and Julius J. Menn
ACS SYMPOSIUM SERIES 255; 351 pp.; ISBN 0-8412-0852-2

"Advances in Pesticide Formulation Technology"
Edited by Herbert B. Scher
ACS SYMPOSIUM SERIES 254; 264 pp.; ISBN 0-8412-0840-9

"Structure/Performance Relationships in Surfactants"
Edited by Milton J. Rosen
ACS SYMPOSIUM SERIES 253; 356 pp.; ISBN 0-8412-0839-5

"Chemistry and Characterization of Coal Macerals"
Edited by Randall E. Winans and John C. Crelling
ACS SYMPOSIUM SERIES 252; 192 pp.; ISBN 0-8412-0838-7

"Conformationally Directed Drug Design:
Peptides and Nucleic Acids as Templates or Targets"
Edited by Julius A. Vida and Maxwell Gordon
ACS SYMPOSIUM SERIES 251; 288 pp.; ISBN 0-8412-0836-0

"Ultrahigh Resolution Chromatography"
Edited by S. Ahuja
ACS SYMPOSIUM SERIES 250; 240 pp.; ISBN 0-8412-0835-2

"Chemistry of Combustion Processes"
Edited by Thompson M. Sloane
ACS SYMPOSIUM SERIES 249; 286 pp.; ISBN 0-8412-0834-4

"Catalytic Materials:
Relationship Between Structure and Reactivity"
Edited by Thaddeus E. Whyte, Jr., Ralph A. Dalla Betta,
Eric G. Derouane, and R. T. K. Baker
ACS SYMPOSIUM SERIES 248; 470 pp.; ISBN 0-8412-0831-X

"Polymer Blends and Composites in Multiphase Systems"
Edited by C. D. Han
ADVANCES IN CHEMISTRY SERIES 206; 400 pp.; ISBN 0-8412-0783-6